Out *of* Print

A VOLUME IN THE SERIES
Page and Screen

EDITED BY
Kate Eichhorn

OUT *of* PRINT

Mediating Information in the Novel and the Book

Julia Panko

UNIVERSITY OF MASSACHUSETTS PRESS
Amherst and Boston

ISBN 978-1-62534-560-8 (paper); 559-2 (hardcover)

Designed by Deste Roosa
Set in Perpetua
Printed and bound by Books International, Inc.

Cover design by Rebecca Neimark, Twenty-Six Letters
Cover art by Su Blackwell, *The Book of the Lost* © 2010. Courtesy of the artist.
www.sublackwell.co.uk

Library of Congress Cataloging-in-Publication Data
A catalog record for this book is available from the Library of Congress.

British Library Cataloguing-in-Publication Data
A catalog record for this book is available from the British Library.

Portions of chapter 3 appeared in a different form as "'A New Form of the Book':
Modernism's Textual Culture and the Microform Moment" in *Book History* 22
(2019): 342–69, copyright © 2019 by Johns Hopkins University Press. Chapters
3 and 4 draw on material published as "'Memory Pressed Flat Into Text': The
Importance of Print in Steven Hall's *The Raw Shark Texts*" in *Contemporary Literature* 52, no. 2 (2011): 264–97, copyright © 2011 by Board of Regents of the University
of Wisconsin Press. Reprinted courtesy of University of Wisconsin Press.

for Séamus

CONTENTS

PREFACE

James Joyce once declared that he wrote *Ulysses* to "give a picture of Dublin so complete that if the city one day suddenly disappeared from the earth it could be reconstructed out of [his] book."[1] This now-famous quote is quintessential Joyce: he was an author who rarely shied away from hyperbole or self-promotion. But the statement is also illuminating for its figuration of the novel as a kind of medium, capable of preserving information about the city. Equally illuminating is Joyce's reference to *Ulysses* not as his "novel" but as his "book." This slide from genre to medium hints at the novel's close and centuries-old association with the book. Joyce's choice of language is all the more significant given that the new media of his day had already begun to replace the print book for information management. If the phrase "the death of the book" feels critically exhausted in the twenty-first century, this is not only because the dominance of digital information media has increased claims of the book's death but also because such claims were already an established trope when Joyce was writing *Ulysses*. Consider the Futurist Manifesto's assertion that "the book, a wholly passéist means of preserving and communicating thought, has for a long time been fated to disappear."[2] Or consider Walter Benjamin's more measured analysis of academic scholarship's movement away from the book. "Today the book is already . . . an outdated mediation between two different filing systems," he wrote in 1928, because "everything that matters is to be found in the card box of the researcher who wrote [the book], and the scholar studying it assimilates it into his own card index."[3] A century later, we might substitute "computer" for "card index." To write about the history of the book—or to imagine its future—is to contend with its perpetual obsolescence. In no realm has this been truer than in the realm of information mediation.

I came to the critical study of information and the book by way of the novel. During a busy semester, I picked up a copy of *House of Leaves*. Mark Z. Danielewski's novel was a welcome reminder of the essential pleasure of reading a deeply engrossing narrative. It also astonished me with its innovative design, disrupting my expectations of what a book—or a novel—should look like. *House of Leaves* is what I refer to in this volume as a metamedial novel: one that self-consciously calls attention to its medium and its mediations, as a metafictional novel foregrounds its fictionality.[4] With its typography twisting and turning across the pages to represent the architecture of the titular house, *House of Leaves* is a testament to the aesthetic and even mimetic potential of the book's textual materiality. It is also a testament to the digital technologies that made possible its composition and publication. Danielewski's bookish tour de force thus sent me outward into the digital landscape as I began to investigate the complex interconnections among information, the novel, and the book. It also sent me back to the modernist period. Attending to the specificity of Joyce's phrasing, I began to consider the bookishness of his scrupulously detailed record of Dublin. Subtly but persistently, *Ulysses* reflects on its existence as a novel printed in a book. This line of inquiry led me to view *Ulysses* as symptomatic of a wider interest in the book among modernist writers. Metamedial experimentation with the form of the book constitutes an understudied aspect of literary modernism's fascination with form. This experimentation emerged in dialogue with broader conversations occurring about information during this touchstone moment of media transition, as information professionals looked from books to card catalogues, files, and microform.

Out of Print is a study of the role the book has played in information's imaginary—a role that encompasses the book's material qualities, affordances, and histories of design and use as well as the affective sway it has held over readers. My analysis centers on the early twentieth and twenty-first centuries, tracing parallels between these two moments when the book's future as an information medium has seemed especially precarious. During both, there was a shift in information management from books to newer media that required the use of interfaces to access information. The highly mediated nature of information accessed in this manner gave rise to the perception that information was essentially immaterial. In contrast, the book's accessible tactility has had the effect of foregrounding its forms. The metamedial novels written during the modernist and contemporary periods bear witness to, and intervene in, debates about form, medi- ation, information, and the book. My analysis of these debates reads information culture via literary culture and vice versa. The death of the novel has frequently been declared in tandem with that of the book, and the ways in which the book mediates information has influenced the novel's longstanding vocation as a cultural archive. Novelists are well positioned to examine how the transition from books to the regime of modern information mediation has produced distinct ideologies about information.

The book and the novel have quietly shaped ideas about information, including how we represent and navigate the world's information and how we strive to collect and preserve information about the places, people, and ideas that matter to us. In developing a critical account of information mediation from the intertwined perspectives of book history and the theory of the novel, I am arguing for the importance of acknowledging the forms that underlie information. From information's instantiation in media to its organization in systems, form has been central to information's meaning to a degree often occluded by newer information media and modern practices of information management. My analysis, in turn, expands the ways in which literary critics can discuss form in the novel, particularly in theorizing the novel's archival ambitions and its experimentation with the book's materiality. The media through which we consume information—and novels, for that matter—have perhaps never been so diverse: print books exist in a vibrant media ecology that includes e-readers, mobile phones, voice-user interfaces, laptop computers, and many other platforms. As user-friendly interfaces enable the passive consumption of information, studying the book and the novel can help us grapple with the forms, and the consequences, of modern information mediation.

ACKNOWLEDGMENTS

I owe a great debt of gratitude to the many people who have supported me in the process of seeing these ideas take their form in print. During the early stages of this project, Alan Liu's astute counsel helped me refine and clarify my claims. Rita Raley's probing questions led me to deeper and richer lines of inquiry. I thank them both for their rigorous engagement and generous mentorship. Enda Duffy reminded me to avoid, above all, being boring; I hope I have succeeded. Michael LaGory helped me hone my prose. Paul Saint-Amour gave me the best possible introduction to *Ulysses*. I am a better scholar and teacher thanks to his model, and this book owes much to his influence.

Discussions with friends and colleagues opened up new avenues in my thinking. I am especially grateful to Kevin Kearney, Lara Rutherford-Morrison, Marthine Satris, Tim Gilmore, Megan Palmer, and Robin Chin Roemer for illuminating conversations at the University of California, Santa Barbara; and to Arthur Bahr, Diana Henderson, James Buzard, Stephanie Frampton, Sandy Alexandre, Nick Montfort, Shankar Raman, Gretchen Henderson, and Kelley Kreitz at the Massachusetts Institute of Technology. As I completed this work at Weber State University, Molly Morin, Jennifer Mitchell, Liese Zahabi, and Michael Wutz provided intellectual camaraderie and insightful feedback for which I have no adequate words of thanks.

At the University of Massachusetts Press, Brian Halley provided clear and cogent guidance at every step. I am grateful for his advice and advocacy. I thank Kate Eichhorn for her valuable suggestions for revising the manuscript, Rachael DeShano for her professionalism and warmth throughout the production process, and Dawn Potter for her keen eye and elegant copyediting. Two anonymous readers for the press provided thoughtful comments that helped me better articulate the manuscript's central argument.

Fellowships from the Mellon Foundation, the University of California Humanities Network, and the Interdisciplinary Humanities

Acknow-
ledgements

Center at the University of California, Santa Barbara, were instrumental to my ability to undertake this research. Weber State University's Department of English, College of Arts and Humanities, and Research, Scholarship, and Professional Growth Committee also provided financial support. Louise O'Connor at the National Library of Ireland and Stéphanie Manfroid at the Mundaneum Archive Center in Belgium deftly assisted me with my research queries. Additionally, I am grateful to the Johns Hopkins University Press and the University of Wisconsin Press for permission to reprint material that appears in chapters 3 and 4. I also thank Craig Saper and the Roving Eye Press, Canongate, and Penguin Random House and Penguin Random House UK for permission to use the figures I have included.

Writing can be a solitary task. I am both lucky and deeply grateful to have had the enthusiastic support of friends and family throughout this endeavor. Thanks especially to David Panko, Liz Melicker, and Kate Clevenger for their encouragement, and to Nora Leonard and Diane Mitchell for the invaluable gift of more time to write. My parents, Rosemarie and Raymond Panko, have been unflagging champions of my work. I am indebted to them for nurturing my curiosity and eagerly reading so many drafts. My children, Declan and Aoibhín, remind me every day of the pleasure books can bring. I thank them for being patient and keeping me grounded as I wrote this one. Finally, Séamus Leonard is my best collaborator. His support of me, and of this project, has been immeasurable. This book is for him.

OUT *of* PRINT

Information beyond the Book

Scale, Mediation, and the Novel since Modernism

In books lies the soul of the whole Past Time; the articulate audible voice of the Past, when the body and material substance of it has altogether vanished like a dream. . . . All that Mankind has done, thought, gained or been: it is lying as in magic preservation in the pages of Books.

—Thomas Carlyle, On Heroes, Hero-Worship, and the
Heroic in History

As long as the book was responsible for all serial data flows, words quivered with sensuality and memory. It was the passion of all reading to hallucinate meaning between lines and letters. . . . Electricity put an end to this. Once memories and dreams, the dead and ghosts, become technically reproducible, readers and writers no longer need the powers of hallucination. Our realm of the dead has withdrawn from the books in which it resided for so long.

—Friedrich Kittler, Gramophone, Film, Typewriter

In a 1927 review, D. H. Lawrence described John Dos Passos's experimental novel *Manhattan Transfer* as a multimodal information storage system: "if you set a blank record revolving to receive all the sounds, and a film-camera going to photograph all the motions of a scattered group of individuals, at the points where they meet and touch in New York, you would more or less get Mr. Dos Passos' method." If art holds a mirror up to nature, Lawrence's review suggests, then the novel holds up audio and visual recorders. For him, the end result of Dos Passos's process is perfectly mimetic representation, where "the book becomes what life is, a stream of different things and different faces rushing along in the consciousness."[1] What might we make of this medial vision of literary representation, where a novel records the world around it like an automated machine? And how might Lawrence's characterization of a novel as an information medium hold up nearly a century later, when storage media have evolved from cameras and gramophones to multi-terabyte hard drives?

To begin unpacking Lawrence's review, we should note that his guiding assumption is a familiar one: that a novel can record the real world. The conceit that literary texts act as cultural archives is longstanding. The dream at the heart of literary representation is preservative: to store a subject in words with absolute fidelity. This literary archival drive has been evident in the novel throughout the genre's history. Walter Scott claimed that his historical novels preserved the fading culture of the Scottish highlands; James Joyce boasted that he wrote *Ulysses* in order to create a record of the city of Dublin. As Ian Watt writes in his seminal account of the novel, "the majority of readers in the last two hundred years have found in the novel the literary form which most closely satisfies their wishes for a close correspondence between life and art."[2] Novelists, scholars, and readers alike have conceived of the novel as an inherently informational genre since its inception. To quote E. M. Forster: "what a lot we learn from *Tom Jones* about the west countryside."[3]

What critical studies of the novel have yet to sufficiently account for, however, is the degree to which this informational drive has been shaped by the media format most closely connected with the novel: the print book. Lawrence's review, for instance, situates *Manhattan Transfer* within a network of other information media, imagining that novels might compete with cameras and gramophones in their ability to record the world. Lawrence figures this novel as an archive whose full range of recording encompasses the capabilities of multiple new media. Yet Dos Passos's novel is also a product of its own medial form. *Manhattan Transfer* creates its jarring effect of "scenes [that] whirl past like snowflakes" and the "breathless confusion of isolated moments," as Lawrence describes it, partly through its unusual typographical layout.[4] Alternating among standard text, the capital letters of newspaper headlines, and italicized phrases, the page becomes a visual field, evoking other information media as it demonstrates the flexibility of the print book to simulate them. If the principal trope in Lawrence's review is relatively common—the novel records the chaotic, fleeting impressions of life in the modern city—the review also provides a subtler insight: that experiments with the textual materiality of the page may reveal how the novel's informational dimensions are conditioned by the print book. It is the book that gives form to narrative, to the novel as a genre, to the printed page, and to the information that is incorporated into the text. It is "the *book*" that "becomes what life is."

The material and conceptual connections between the novel and the book, although never absolute or essential, have been remarkably persistent.[5] These connections have become fraught in the twenty-first century, as the sheer scale of digital media, combined with algorithmic computation's purported ability to explain the world through data analysis, threatens to render both books and novels obsolete. In this volume, I argue that the form of the book, in its complex co-evolution with media and information forms, has had a substantial impact on the novel's role as an information medium—and that the novel, in turn, offers crucial insights into the consequences of the book's marginalization in information management. What is lost in today's era of information scale, given that people must rely on automated systems and mediating interfaces to navigate data, is knowledge of the ways in which information is organized and embodied by media. I trace the roots of this contemporary obscuration of form to the early decades of the twentieth century, studying how the print book fared as the burgeoning profession of information science grappled with unprecedented quantities of data. Pairing close readings of novels that emphasize their informational qualities with case studies from information history, I argue that, for novelists and the reading public as well as for information professionals, the print book has served as a counterpoint to conceptualizations of information as immaterial, ineffable, and formless. As the novels I study demonstrate, if the phrase "out of print" signals the possible obsolescence of the book, it also indicates the medium's aesthetic potential—what can be created and formed out of print.

Out of Print makes three important interventions in debates about how books and novels mediate information. First, I analyze the early twentieth and twenty-first centuries as parallel moments of media transition, studying how, in both of these periods, explosions in the scale of information resulted in the marginalization of the print book. I argue that the development of systematic information management was predicated on a shift from the book to newer, interface-based media and that this shift created a model of mediation premised on obscuring the aesthetics of information—the forms that make it meaningful. Put another way, this transparent model of meditation, in which users need not be aware of how information is organized or processed within larger systems, holds a central place in what I will refer to as modern information management.[6] This model is the cause and the consequence of a programmatic movement away from the book. Examining these changes, I chart how the book has functioned as a touchstone for conversations about information scale and its management during two of the past century's most intense media transitions.

The print book has become a residual medium. Raymond Williams defines the *residual*, which he distinguishes from the *dominant* and the *emergent*, as that which "has been effectively formed in the past, but is still active in the cultural process" and "which may have an alternative or even oppositional relation to the dominant culture."[7] Residual media are not entirely displaced; they persist, albeit in a diminished or altered capacity. One reason the book persists is its ability to communicate information's formal structures with an immediate apprehensibility. Whereas the designers of algorithms and search interfaces adopt the metaphors of the inscrutable black box and the opaque cloud, many of the book's metaphors instead invoke legibility: an open book, reading someone like a book, and so on. I use Jay David Bolter and Richard Grusin's term *hypermediacy* to describe how the book's formal elements (including its materiality, tactility, aesthetics, and organization), history of use, and other qualities have meant that it has tended to "remind the viewer of the medium."[8] Laying bare its own processes of mediation, the book's affordance of hypermediacy contrasts with the ideal of transparent mediation.[9] The book's hypermediacy has become a more salient feature in comparison to newer media over the course of the past century. "The book" is not a monolithic subject, of course; it encompasses many registers, from the structures of page, print, and codex to its cultural cachet. My research shows how, in all of these registers, the book has contributed to information's imaginary.

My second intervention moves from medium to genre, establishing the novel as a key site for understanding the book's impact on the representation and mediation of information at scale. I argue that information management's pivot away from the book has had a profound effect on the novel. Since the modernist period, I contend, the novel's archival project has taken a critical turn, from realism to an investigation of the structures, affordances, and ideologies of information media—an investigation channeled through experiments with the form of the print book. Influenced by N. Katherine Hayles's foundational approach, much work in what we might call literary media studies has examined literary texts' self-conscious mobilization of the media that embody them.[10] While novels have routinely served as case studies, this critical literature has rarely focused on formal play with print and page as an issue of genre. I analyze why emphasis on the book's form becomes prominent in novels during the modernist and contemporary eras, a pairing I will unpack momentarily. I examine a range of experimental novels, each of which foregrounds its textual materiality and incorporates the aesthetics of other information media in order to address questions about information scale and its mediation.[11] Some of these, such as James Joyce's *Ulysses* and Mark Z.

Danielewski's *Only Revolutions,* use encyclopedic form to consider how the arrangement and contextualization of information produce its meaning; some, such as Virginia Woolf's *Orlando* and Jonathan Safran Foer's *Extremely Loud and Incredibly Close,* highlight print's tactile properties to explore the implications of information's embodiment in different media. As these novels contemplate the ways in which books limit, organize, and embody text, they insist that data only become meaningful information insofar as they are given form. These novels challenge and critique modern information management's prevailing ethos of transparent mediation.

The subject of my analysis is thus neither solely the novel nor solely the book, but both. This doubled focus is necessary because the novel and the book are interconnected in deep and complex ways, particularly in the realm of information. I take a medium-specific approach to the novel, writing the genre's medial history into its theorization as a genre.[12] I also take a genre-specific approach to the medium, documenting the novel's influence on ideas about the book. Novels combine long-standing generic investments in information, books, and mediation. (I suspect this is why they have proven compelling to media theorists.) Kate Marshall writes of modernist novels: "Observation machines otherwise known as novels have an affinity for testing and staging their own mediality."[13] The novel's archival drive lends it a special propensity for assimilating other informational genres, and the novels I examine all position novelistic narrative as explicitly informational—as conveying information about the world and as adopting the forms of information media. The novel's centuries-old association with the print book, in turn, has shaped its narrative form.[14] Novelists therefore may confront mediation on several fronts: the book's storing of information, the book's organization of literary text, and narrative's incorporation of information. A novel may seem like an archive or a medium, as Lawrence suggests; but novels are also meta-archives and meta-media, forcing readers to rethink the nature of information and form.

In *Out of Print,* I work not only to illuminate undertheorized aspects of the history of the novel, the cultural impact of the book, and the consequences of modern information management but also to demonstrate that issues of form go to the heart of all three. My third intervention is to track these issues across time. I argue that the logical foundations of contemporary information management originated in the early twentieth century, as did the novelistic response of mobilizing the book's form to expose unexamined assumptions about information's mediation. During the first decades of the twentieth century, the rapid expansion of information media sparked a crisis of scale. To address these conditions, modern information management was founded on the principle that information should be managed through interfaces. These interfaces, both physical and processural, made information at once highly structured and highly mediated. Information managed in this manner appeared to be neutral. These changes explain why novelistic emphasis on the book's form, which dates back at least as far as Laurence Sterne's *Tristram Shandy,* intensified during the early twentieth and twenty-first centuries. Such experimentation may be found in novels before and between these two eras, but it is a dominant feature in modernist and contemporary novels. I argue that these moments of concerted novelistic engagement with print textuality have occurred in response to similar discussions about the management of new magnitudes of information—discussions that have frequently invoked the death of the book. The twenty-first century may be the post-print era, but the modernist period was already becoming post-book.

My aim in what follows is not to compile an exhaustive history of novelistic experimentation, information mediation, or debates about the book's future. Nor do I intend to write a defense of (or nostalgic elegy for) the novel or the book. I employ a comparative framework in order to construct one genealogy of twenty-first-century information culture, a strategic account that recontextualizes current understandings of the print book's role in mediation and invites a reconsideration of the ways in which terms such as *information, novel, book,* and *form* have been used in the past century. Scholars affiliated with the new formalism have investigated why forms obsolesce or reactivate at different points in history—what Caroline Levine refers to as "the *longue durées* of different forms, their portability across time and space."[15] My research offers a model for how this inquiry might proceed when focused on the mutual imbrication of genre and medium. Ultimately, *Out of Print* explores the ways the novel and the book may provide insight into modern information mediation by calling attention to form. Information is emergent, becoming; it is produced when data are made meaningful by the forms they take as they are recorded, stored, and circulated. Too often, awareness of these forms becomes a casualty of modern information management's efficient mediation, with far-reaching social consequences. By studying print books—and the novels that emphasize them—we can read form back into information.

FICTIONAL ARCHIVE: THE NOVEL AS INFORMATION

As my opening remarks on the novel's archival impulse indicate, writers as well as literary theorists have consistently regarded the novel as a genre with a special stake in information. In this view, the novel's expansive and detailed accounting of the world produces not just a representation but a record. The notion that the novel acts as an archive, even a reference book, recording information about the time and place it describes, has been persistent enough that critics identify subgenres such as "archival novels" and "encyclopedic novels."[16] Seen in this light, novels like *Ulysses* may be unusual in the degree to which they highlight their archival and medial dimensions, but they are symptomatic of a much broader generic investment in information.

There are several related but distinct reasons why the novel has garnered its reputation an informational genre. First, because novels are typically written in prose, they occupy a discursive middle space between writing that is overtly marked as literary and writing that is not. How to define the literary is a question I take up in chapter 2; for now, I note the critical tendency to define it as writing that has qualities beyond the "merely" communicative or informational—writing that, as Arthur Bahr puts it, is defined by its "excess," its "refusal to submit to the denotative."[17] Although novels use literary techniques and devices that would be lost in a summary of the plot, paraphrasing a novel seems to be a lesser transgression than paraphrasing poetry is. (Tellingly, Cleanth Brooks's examples of the heresy of paraphrase are all poems.) If "the fundamental aspect of the novel is its story-telling aspect," as Forster argues, it follows that novels would be less invested in emphasizing their materiality because "informational texts seek to minimize their perceptual features in the belief that texts calling attention to their vehicular forms interfere with the transmission of their ideas."[18] The convention (however problematic or reductive) that novels are read primarily for their content aligns them as much with informational texts as with literary ones.

The claim that novels contain information is most often derived from the truism that they capture something real despite the fictional status of their narratives. The novel functions like an archive, this argument goes, because its narrative content bears a close relationship to real life. Watt's view on this matter has been foundational: "There are important differences in the degree to which different literary forms imitate reality; and the formal realism of the novel allows a more immediate imitation of individual experience set in its temporal and spatial environment than do other literary forms. Consequently the novel's conventions make much smaller demands on the audience than do most literary conventions."[19] While realism is far from the only mode in which the novel operates, it has been fundamental to critical definitions of the genre. J. Paul Hunter writes that "realism is a relative matter, but in discussions of the novel, the term has tended to become normative, so that novels tend to be judged qualitatively on the degree or amount of realism to be found in each, as if more is better."[20] The novel's realism has taken many forms, from historical

9

realism (the accumulation of details registering the habits of daily life in a certain time and location) to cognitive realism (the realistic representation of thought).[21] Scholarship on the novel consistently characterizes it as the genre of the nation, or of modern subjectivity; whether one views a particular novel as realistically capturing the workings of one mind or the quotidian details of life in a particular place, the assumption is that there is a stable referent that the novel represents. This is the paradox of the novel's archival nature: even when describing a fictional referent—the mind of Leopold Bloom; the social milieu of Casterbridge, Wessex—a novel nonetheless records some real-world truth that this fictional subject models.

A focus on realism, however, does little to explain why novelists began to move away from this style at the same time that information proliferation and its management became major issues. Although we might say that the novel has always incorporated information in some manner, novelists since the modernist era have undertaken an archival agenda more indebted to the forms and genres of information media than to those of realism. Where an early novel reader might have encountered information about the world (in the guise of "the real") via a set of fictional letters or journal entries, readers of modernist and contemporary novels discover narrative texts that mimic paper file layouts or computer code. The chief questions for the novel in the age of modern information management become: What *is* information? What are the discursive norms that shape its articulation? How might the novel compare to other information media in recording and preserving information? This last question is pressing because many critical accounts of modern information media cast them as antagonists to literary representation, rehearsing the Kittlerian argument (encapsulated in the epigraph to this chapter) that literature's once-unique ability to store sensory data was imperiled by the introduction of recording technologies.[22] Yet the capacity for direct representation is not the only vector by which literary works may compete with information media. Modernist and contemporary novelists have made the case for the genre's relevance by grounding their engagement with these questions in an investigation of the forms and affordances of the print book.

Before we proceed further, a few definitions are in order. What is information, and how does it differ from data or knowledge? The smallest, most discrete entity of the three is data, which we might think of as "units or morsels of information."[23] Where a fact is self-evident and true, a datum is a given, a measurement or a rhetorical statement: "data are representations of observations, objects, or other entities used as evidence of phenomena for the purposes of research or scholarship."[24] Used in the aggregate, data's explanatory force accrues collectively. On the other end of the spectrum lies knowledge. From the partial to the total: if data are "units or morsels," knowledge is "the sum of what is known."[25] Of data, information, and knowledge, knowledge is the closest to what we would categorize as wisdom or truth. It is also the most subjective category, to the extent that knowledge requires "an individual knower."[26]

Between data and knowledge lies information, a notoriously difficult word to pin down. Its meaning is often taken as self-evident. Definitions vary from the broad ("any given [datum] of our cognitive experience that can be materially encoded for the purpose of transmission or storage"), to the technical ("a mathematically defined quantity . . . which represents the degree of choice exercised in the selection or formation of one particular symbol . . . out of a number of possible ones, and which is defined logarithmically in terms of the statistical probabilities of occurrence of the symbol"), to the cryptic ("a difference which makes a difference").[27] Here I settle on a working definition: information is data made meaningful by their contextualization. Information, in other words, is meaningful because it takes individual data points and organizes them or otherwise gives them form, setting them in relation to one another. Form lies at the heart of information. This act of giving form to data is what allows for information's interpretive power. Whereas data are abstract, information is necessarily formed and articulated. There is still an autonomy to this formal organization; while knowledge requires a knower, information does not.[28]

As a result, *information* is the term that hews most neatly to the category of fiction. A novel is a fictional work; its epistemological status differs from that of discourses such as scientific writing that purport to objectively describe the factual state of the world. Yet a novel may integrate information about the world, processing this information by embedding it within its narrative. Some novels even include lists of facts or other data structures. For the most part, however, a novel is less likely to contain facts—declarative statements about the true state of reality—than that fuzzier, more interpretive and rhetorical category of information. "The novel," wrote Forster, "whatever else it may be, is partly a notice board"; he characterized it also as the genre in which "information abounds."[29] This statement implies not only that novels are archives of the cultures that produce them but also that the novel's archival drive is specifically informational. It is not simply that a novel contains an impression or representation of reality but that it contains specific details, meaningfully arranged, in forms that are similar to nonfictional, nonliterary informational texts.

This explains why critical descriptions of novelistic form have been so vexed. The genre's theorists consistently disagree on its defining formal characteristics—or whether it even has any. To cite a few canonical voices: Watt describes "what is often felt as the formlessness of the novel, as compared, say, with tragedy or the ode." Erich Auerbach speaks of "the broad and elastic form of the novel," and, for Mikhail Bakhtin, the novel is "plasticity itself," always "subjecting its established forms for review." Woolf describes the novel as the "most pliable of all forms." More recent theorizations make similar claims. To Terry Eagleton, the novel is "the most hybrid of literary forms"; in Michael McKeon's view, the novel "self-consciously incorporat[es], as part of its form, the problem of its own categorical status."[30] If theorists of the novel agree on any point, it is that this genre is definable only through its lack of stable formal attributes.

Rather than an absence of form, however, the issue is an overabundance of forms. If the novel is difficult to define, this is a consequence of its expansiveness. Since its inception, it has habitually drawn on a range of genres and discourses, including informational ones. Early novels appropriated different styles of writing, especially nonfictional genres such as travelogue (Daniel Defoe's *Robinson Crusoe* and Jonathan Swift's *Gulliver's Travels*) and epistolary correspondence (Samuel Richardson's *Pamela* and *Clarissa*). It was commonplace to use the word *history* in titles (as in *The History of Tom Jones, a Foundling*). Early novelists and readers defined the novel in comparison to other discourses that blurred the line between fact and fiction, including journalism.[31] Given the practice of imitating and incorporating nonfictional genres, it is not a great stretch for a novel to imitate explicitly informational texts such as encyclopedias or Internet news feeds (examples I discuss in chapters 1 and 2). Formal flexibility is one reason the genre was named for novelty: its push for newness has consistently come through stylistic experimentation with the conventions of other discourses.

At the macro level, then, the novel's integration of informational discourses is part of its tradition of formal flexibility. At the micro level—the level of the individual page, sentence, and word—novels have incorporated content and textual structures we would not ordinarily consider literary. A double-entry bookkeeping log, a long list of items sitting on a kitchen shelf: such examples go beyond extending the verisimilitude of the narrative world.[32] They interrupt the narration with a glut of information, momentarily turning the novel into a reference book. To examine the novel's informational qualities is to address questions about the genre as a whole: how it is defined (and defies definitions), how it does or does not employ literary language, and how it engages with information as content and as form.

MEDIATING SCALE: INFORMATION AND THE NOVEL IN THE DIGITAL AGE

In *Out of Print,* I analyze novels that explore the medial underpinnings of this long-standing generic investment in information. These texts consider how their subjects are doubly mediated, first as narrative representations and second as text embedded in books. While their narratives contend with the task of representing information about places, objects, and people, their unconventional uses of the print book reveal the book's storage capacities and limitations. This strategy emerged as a literary response to a paradigm shift in information culture. The reason that self-reflexively "bookish" novels (to use Jessica Pressman's terminology) flourished during the early twentieth century, I am arguing, and the reason they are flourishing again today, is that these are threshold moments in which the scale and mediation of information reached a critical juncture.[33] The novelists I study resist modern regimes of mediation, from the systematic management of the modernist period to the corporatization of information under Big Data.

Information abundance is a defining feature of contemporary information culture. (I will discuss information proliferation during the modernist era at the end of this chapter.) In the twenty-first century, human communication creates 5 billion gigabytes of information every two days.[34] Digital devices capable of producing and storing historically unprecedented volumes of information are omnipresent, and information overload is enough of a cliché to have spawned its own Internet vernacular—"TMI," "TL;DR." Contemporary information culture is also distinguished by the mediation of scale through technological management. Whether through a command line, a graphical user interface, or one of the so-called invisible interfaces of ubiquitous computing, the user always interacts with a mediating system.[35]

More recently, information mediation encompasses the algorithmic analysis of large data sets. This capacity underlies Big Data, the phenomenon that exemplifies how thoroughly scale and its mediation have transformed people's relationship to information. Although the term entered common usage only in the second decade of the twenty-first century, Big Data quickly became credited with changing the nature of what could be investigated and known.[36] "At scale," writes Christine Borgman, "big data make new questions possible and thinkable. For the first time, scholars can ask questions of datasets where $n =$ all."[37] According to danah boyd and Kate Crawford, "the widespread belief that large data sets offer a higher form of intelligence and knowledge that can generate insights that were previously impossible" has become akin to "mythology."[38]

Big Data is best understood as a matter of relative rather than absolute scale—"data of a very large size, typically *to the extent that its manipulation and management present significant logistical challenges.*"[39] Big Data is too much data to be able to manage or use without the intervention of sophisticated systems. Even standard software falls short of the task. Big Data is mediated: without algorithms, sophisticated processing software, and other management systems, the data cannot be usefully analyzed. Big Data, then, is not a set quantity of data so much as a set of attributes or assumptions about what can be done with large, complex data sets. It encompasses the physical forms that data take and the technologies used to analyze them as well as ideas about data: the implications of scale; the ways in which technology mediates our perception of data; the distinction between raw data and meaningful information; and the explanatory power of Big Data. If overload is one side of the information abundance coin, Big Data is the other, for it promises that otherwise-excessive collections of data may be managed. As information is managed by complex systems, however, users cannot directly view either its organizational form or its material form. Information abundance dictates that when we attempt to encounter information at scale, we do so without trying to actually comprehend that scale, without grasping what all of the data mean as a gestalt. If we access individual pieces of a data set, we access them as fragments with no meaningful relation to the whole;

more likely, we allow search, analysis, navigation, or visualization programs to analyze the sum of the data and return results and conclusions. This is a useful process, but it is distinct from understanding the information in and of itself.

The conditions I am describing—information scale, the dominance of digital media, and technological mediation—would seem to make obsolete the novel's ability to record information. Indeed, the idea that the contemporary information ecology is antagonistic to the novel surfaces frequently in discussions regarding the transition from print to digital media. Digital media, one argument goes, alter human cognition by privileging hyper attention over deep attention—the latter being necessary for novels, whose length and complexity necessitate prolonged, intensive reading.[40] Without wishing to reify this view, I note how commonly scholarship on the attentional impact of digital media has used the novel as a reference point. Nicholas Carr and Sven Birkerts cite novels by authors including Charles Dickens and Henry James as exemplars of texts that require the sustained, contemplative reading they fear the Internet is destroying. Hayles similarly associates deep attention with this genre: her examples of activities that require deep attention include reading "a novel by Dickens" and Jane Austen's *Pride and Prejudice*.[41] While accounts of digital media's effects on cognition range from measured to alarmist, such concerns speak to a belief that the novel is linked to a print-centric reading culture and that it is at odds with digital media as a result.[42]

Another line of argument views the novel as inherently opposed to digital information. The most obvious contrast is one of scale. The novel is the literary genre known for, and typically defined by, size. In Forster's formulation, for example, a novel is "any fictitious prose work over 50,000 words." Naomi Baron notes that "the novel as a literary genre entered English with a weighty presence"; she cites Penguin Classics editions of Richardson's *Pamela* and *Clarissa* as being 544 pages and 1,536 pages, respectively.[43] The tendency to conflate the novel with the print book likely has to do with the fact that the former could not have flourished without the latter because of the genre's noticeable length. As Walter Benjamin wrote, the novel is "distinguishe[d] . . . from the story (and from the epic . . .) in its essential dependence on the book" because "the dissemination of the novel became possible only with the invention of printing."[44] To produce and circulate numerous copies of a long text, a publisher needed the mass-production technology of movable type and preferably the codex format to organize the pages. But a single novel that is very long—or even serialized, such as Danielewski's multivolume *The Familiar* (whose first five of twenty-seven planned volumes run to more than 4,000 pages)—is dwarfed in scale by digital information storage. The amount of data needed for Big Data analysis is at an even greater degree of remove. When novels attempt to integrate information into their narratives, that information must necessarily be of a far lesser quantity than can be stored by newer media.

Other theorists have argued that the novel is (or appears to be) obsolete in the age of digital information because computational methods of storing and processing information differ profoundly from literary narrative in their ability to represent and make sense of the world. Lev Manovich has described narrative and database as opposing cultural forms, arguing that database has usurped narrative's cultural position. While his characterization of these two forms as antithetical (as "natural enemies") has detractors, even accounts that characterize narrative and database as "natural symbionts" confirm their vastly different modes of operation.[45] Manovich's argument is symptomatic of the view that the culture of digital information is at odds with the novel because of how each approaches explaining the world. Michael Wutz writes that "the novel . . . can yet insist on a notion of information, and on the processes of commuting such information into such . . . terms as *knowledge, insight,* and *wisdom,* that are qualitatively different from the binary bits of computer processing," so that novels can "offer a system of information and knowledge self-consciously different from the computable databases of mainframes and networks."[46]

In contrast, the logic of Big Data—and of the information scale and mediation practices for which it often stands as synecdoche—insists that data analysis explains the world by finding patterns not previously discernable. This is not knowledge or insight but pattern recognition through statistical analysis. If narrative used to be the dominant system for making sense of the world, Big Data, rather than database, is now its clearest challenger. In all of these critical narratives, scholars invoke the novel's tie to the print book, implicitly or directly: they assume that, as the book becomes a residual medium, the novel, in turn, becomes a residual form.

METAMEDIA: THE NOVEL AND THE PRINT BOOK

The novels I study contest these intertwined crises of modern mediation: the obscuration of information's meaningful forms and the apparent displacement of novels and books. As we have seen with the example of *Manhattan Transfer,* these texts theorize mediation via their own materiality, disrupting the conventions of page and print. Medially engaged literary experimentation—described variously as "technotext," "multimodal," and "metamedial"—has become a prevalent feature of the contemporary novel (although I argue later in this chapter that its origins lie in the modernist period).[47] The best-known example is Danielewski's *House of Leaves* (2000), a novel whose innovative typographical layouts mimic the disorienting architecture of the eponymous house. Other examples include novels featuring blacked-out sections of text, typographical collages culminating in a flipbook, and the simulation of an old, annotated library book.[48] These novels could not be translated to a digital screen without losing significant aspects of their composition. "If," writes Leah Price, "the book has been invisible (or intangible) to most twentieth-century literary critics, it isn't simply because we aren't trained to analyze material culture; it's also because a commonsense Cartesianism teaches us to filter out the look, the feel, the smell of the printed page." Novels that foreground their textual materiality push back against this invisibility, asserting what Kiene Brillenburg Wurth calls "book presence." Wurth defines book presence as "the effect of an ongoing process of the becoming obsolescent of the book," wherein "the book precisely materialized when it became immaterial as an information medium."[49] Such novels highlight how books function as information media and as literary media.

To refer to this subset of novels, I follow Alexander Starre in using the term *metamedial*. In the tradition of metafiction, metamedial novels self-reflexively call attention to the artifice of narrative, representation, and literary language. The distinction is this: in metamedial novels, linguistic artificiality is conditioned and compounded by the artifactuality of the print book. Drawing on Patricia Waugh's definition of *metafiction*, Starre describes metamediality as "a form of artistic self-reference that systematically mirrors, addresses, or interrogates the material properties of its medium. Literary metamediality therefore draws attention to the status of texts as medial artefacts and examines the relationship between text and book."[50] Metamedial works constitute a specific subcategory of metafiction, in which play with the work's material properties forces the reader to consider the work's fictionality. Studies of metafiction have long been concerned with literary representation. The metamedial novels I examine in *Out of Print* continue this interrogation into the nature of what, exactly, fiction may describe and encode, but they do so by considering how the novel's representative qualities dovetail with or diverge from other methods of recording information, as well as how such recording is shaped by media. My use of the term *metamedial* also evokes the desire evident in these novels to turn books into multimedia meta-media—media capable of subsuming all other media.

Metamedial novels reveal what mediation obscures. We might think of novels as alternative media, leveraging the print book to critique information management's dominant assumptions about form and mediation.[51] In contrast to storage media such as microfilm and computer hard drives, books contain text that is humanly readable. In technological terms, print books operate very differently from interface-based media. At its broadest level of meaning, an *interface* is any "surface lying between two portions of matter or space, and forming their common boundary." My use of the term aligns most closely to the concept of a user interface.[52] I am particularly concerned with interfaces—whether physical objects or processes—that intercede between users seeking information and the information as it is inscribed in the storage medium. Reading screens fall in this category; so do the protocols that govern information retrieval from complex systems, whether these be algorithms or guidelines for information professionals.

Such interfaces are what chiefly distinguish what I am calling modern information mediation from earlier systems of information management. Like the messenger god Hermes, a medium is both an information vessel and an emissary: it carries its content across space (radio, telephone), time (writing, phonography), or both. Mediation entails a rupture between the creation of the message and the moment or place of its reception. Modern information culture is marked by an even greater degree of removal because of the expectation that information has reached a threshold of scale that can only be managed indirectly. The Internet seems nearly infinite, so we locate what we need through search interfaces; a century ago, when there seemed to be too many books, information professionals began to store data on microform media, which they accessed with reading screens. During the past century, the use of such interfaces has encouraged the perception that information is immaterial and formless, an idea to be consumed rather than data instantiated in a medium.

Viewed in this light, descriptions of the page (or book, or narrative) as an interface gloss over the major differences between words printed on a page and bound in a codex and words stored on a hard drive and read on a screen.[53] The former are immediately accessible, combining the site of storage with the site of inscription. The latter, as Lori Emerson describes, by design involve "an interface that recedes from view, ideally to the point of invisibility."[54] Additionally, the volumetric space of the book, like the two-dimensional space of the page, becomes a framework for understanding the organization of its contents. In these ways, the affordances of the print book work against the principles of transparent mediation so integral to modern information management. My research shows that the value and function of media that feature interfaces has consistently been articulated in direct opposition to that of the book.

As metamedial novels reflect on the print book as an object— material, aesthetic, tactile, structured—they expose and challenge the practices and ideas that have underpinned modern mediation. They also prompt a reexamination of the long-standing relationship between the novel and the book. These texts are thus useful for addressing the interoperation of genre and medium. When scholars have discussed media from the perspective of genre theory, they have tended to view the novel in very general media terms ("writing," "written narrative") or to passingly mention that novels are a "print genre" in the sense of being artifacts of predigital or post-oral culture. Even monographs in this field that study the novel's relationship to media rarely discuss metamedial novels or the print book's influence on novelistic form.[55] Conversely, scholars who take a media studies approach to literature have tended to frame their analysis in broad terms—"book fictions" or "experimental literature," to give two examples—treating genre as largely incidental to the works' textual strategies.[56] As a result, while Marshall, Wutz, John Lurz, and others have advanced the theorization of the impact that print and the book have had on the novel, this area remains understudied.[57]

Because metamedial novels foreground this impact, they reveal the usefulness of format as a critical category for the study of the novel and the study of the book. As Jonathan Sterne describes, the category of format is complementary to but distinct from that of medium: "*format* denotes a whole range of decisions that affect the look, feel, experience, and workings of a medium. It also names a set of rules according to which a technology can operate."[58] The format Sterne focuses on, the MP3, is a much more compact field of study than is the print book; the former is tightly delimited, technologically and historically. Moreover, we might devote attention to different formats of print books—hardcover, mass-market paperback, trade paperback, and so on. Yet the category of the print book already offers a strategic condensation and combination of media: for instance, print, type, paper, codex, text. A print book is a mini-media ecology. Its format encompasses physical characteristics (its tactile and bound form), organizational systems (the arrangement of pages within a codex and text within those pages), and phenomenological registers (including aesthetics, sensory attributes, and perceptions of aura). Metamedial novels emphasize these aspects, which are attributes of all print novels and which contribute to the processes by which novels make visible the forms of information.

Format is also significant in its semantic adjacency to genre. Both "novel" and "print book" suggest defaults and norms. Formats provide a valuable framework for theorizing defaults because they are not reducible to either their nondiscursive technical properties or the phenomenological domain of their perception. As Amaranth Borsuk writes, "the thing we picture when someone says 'book' is an *idea* as much as an *object*."[59] A print book codifies a series of default forms, arrangements, and uses, but its affordances also entail potentiality. Defaults are less interesting in and of themselves than as the standards from which works may deviate. So, too, in the case of novels. Novels, more than poetry, correspond to Johanna Drucker's definition of the "unmarked" text, "the single grey block of undisturbed text" that allows readers to read words on the page without seeing them as visual or material artifacts: "Literary works . . . essentially adopted the unmarked mode of Gutenberg's biblical setting as their norm. . . . The aspirations of typographers serving the literary muse are to make the text as uniform, as neutral, as accessible and seamless as possible."[60] Metamedial novels work against this default aesthetic, disrupting expectations for genre and format alike as they adopt the forms of information and information management. Through these novels, we better understand the aesthetics of information scale.

The issue of defaults and deviations brings me to a final note regarding my rationale for focusing on metamedial novels. To what degree can one generalize about the genre from these idiosyncratic texts? Because metamedial novels are highly experimental, they are necessarily limit cases that test the boundaries of the genre. Yet this very quality makes them exemplars of a genre that has been formed through limit-case experiments. The novel is always already a genre of exception. I will have more to say about exemplarity and novelty in later chapters; here, I propose that, while the self-conscious book-ishness of metamedial novels is unusual, it is indicative of the ways in which the novel's encounters with information have been shaped by the book throughout its history. Metamedial novels not only reveal the conventions of novels and of books; they also remind us that the relationship between the two constitutes its own default.

The novel has never been entirely bound to the book. Historically, many novels were serialized in periodicals, and the connection between novel and book is tenuous in the age of digital e-readers. Novelists are adapting their genre for new platforms such as smartphones, and even print novels are never really "born print" in the manner that digital works are born digital. Today, novels are likely to be born as digital documents.[61] In the past, they would have been born as man-uscript or typescript. The relationship between the novel and the media forms it takes is not reducible to a print-digital dichotomy. Yet the association between novels and books persists. This association is present in the slippage when a colleague says, "I just read a new book on my Kindle." It is present in Robert Coover's argument in his 1992 essay "The End of Books" that hypertext literature would kill the novel by killing the book.[62] It is present when we perform a Google Image search for "novel" and the algorithm returns pictures of books. The connection between novel and book may not be abso-lute, but it is a tenacious and important convention. I map a range of literary responses to information and its mediation, from canonically informational novels such as *Ulysses* to novels whose investment in modern information culture is less obvious. This analysis explores a wide range of potentiality—what the book *may* do for the novel and for information storage, rather than what it must do or what it habitually does.

INFORMATION CULTURE CIRCA 1900: MODERN MEDIATION AND METAMEDIAL MODERNISM

The preceding sections have primarily discussed the media conditions and literary responses of the twenty-first century: algorithmic analysis, digital information storage, recent novelistic experimentation, and so forth. In this volume, I put these twenty-first-century literary and informational landscapes into dialogue with those of the modernist era. Contemporary information mediation did not spring fully formed from digital media. It represents one stage in a history of ideas about how information should be managed and what role the book should play in that management. The excavation of digital media's predigital roots is a widespread practice within media studies, exemplified by work such as Sterne's and Lisa Gitelman's, and many media theorists have identified aspects of pre-1900 media culture that anticipate digital culture. I focus on the modernist period because the first decades of the twentieth century ushered in information management techniques, and novelistic responses to the assumptions entailed by those techniques, that remain active a century later. Information mediation and the future of the book were central issues for both modernism and modernity.

The information culture and media-centered literary experiments of the modernist period share commonalities with those of the twenty-first century that make their comparison especially fruitful. That media were essential to modernism is by now well established. Scholars in modernist studies have examined everything from telegraphs and phonographs to corridors and vulcanized rubber under the rubric of media.[63] Media theorists, too, have cited the importance of the decades circa 1900, building on Kittler's description of the epistemic break ushered in with the discourse network of 1900.[64] (That this media transition was intertwined with modernist literature is evident in Kittler's inclusion of Gertrude Stein as an emblematic figure of that discourse network.) The new modernist studies have asserted that aspects of modernity and modernism extended beyond the temporal and geographical boundaries traditionally associated with these terms.[65] Although I do not claim the contemporary period as precisely modernist, my research contributes to the vertical expansion of modernist studies in its discussion of how and why elements of modernism's information culture have been remobilized a century later.

Additionally, the idea that the links between the early twentieth and twenty-first centuries are worthy of critical attention has been gaining momentum. Manovich, for instance, describes new media's formal and logical indebtedness to the avant-garde cinema of the 1920s. Alan Liu cites Frederick Winslow Taylor's scientific management as an origin point for the structured and encoded discourse that Liu argues is integral to "discourse network 2000." Paul Stephens's history of poetic interventions into the cultural conditions of information overload moves from Stein to contemporary conceptual writing. Jessica Pressman has made the most sustained analysis of the connections between these periods: she argues that the modernist imperative to "make it new" was influenced by media and that it provided a model of experimentation taken up a century later by writers of electronic literature.[66] I uncover further significant connections by demonstrating how, in both eras, concerns about the scale of information, and consequent uncertainty about the print book's future, have been met by renewed novelistic interest in textual materiality as a way of rethinking emerging practices of mediation.

Key features of information proliferation and management that have reached their apotheosis in the digital age originated in the early twentieth century. As chapter 1 describes in detail, while anxiety about information overload predates the early twentieth century, what was new were the forms information took as it proliferated across new media and as it was managed with new techniques. The volume of printed matter exploded. Book production quadrupled in the United Kingdom between the 1840s and 1916. In the United States, the number of new titles being published in 1910 was six times greater than it had been three decades earlier.[67] The claim that "readers [were being] overwhelmed with the avalanche of books," as the *Athenaeum* declared in 1912, became a common enough trope for a columnist in the *Literary Digest* to wearily write in 1923, "There is 'an avalanche of books,' of course."[68] Similar complaints surfaced about other media, as periodical circulation increased dramatically, alongside advertisements, documents, and files. Benjamin worried that children were so besieged by "locust swarms of print" from newspapers, advertisements, and other texts that they lacked the ability to focus on a print book: "Before a child of our time finds his way clear to opening a book, his eyes have been exposed to such a blizzard of changing, colorful, conflicting letters that the chances of his penetrating the archaic stillness of the book are slight."[69] Concerns about information proliferation, and the perception that its scale was historically unprecedented, led to the development of systematic management. The growing cadre of information professionals devised standardized processes for storing and accessing information, positioning organizational systems as interfaces that mediated between users and data. Some information professionals also advocated for the use of microform, which necessitated magnification screens or other reading interfaces. As information became more heavily mediated in these ways, the print book seemed a less efficient storage medium by comparison. Ideologically if not technologically, the information culture of the early twentieth century set into motion practices and assumptions about information and its mediation that continue to operate in the age of Big Data.

Wrestling with the question of whether the novel could meaningfully act as an archive for societies whose informational output had reached watershed levels, novelists looked to the book as a microcosm of information management. Most literary criticism treats metamedial experimentation as a product of the digital era. Hayles and others have noted that this experimentation has intensified in the contemporary novel partly in reaction to the perceived obsolescence of the book in the age of digital information and e-readers and partly because its creation is enabled by the use of digital platforms that allow novelists to more easily design visually complex works.[70] Roughly a century ago, however, metamedial experimentation in the novel also emerged as a trend, in a similar effort to reevaluate the genre's relevance. Marshall has documented the novel's awareness of its function as a medium during this period: arguing that novels like *Manhattan Transfer* and *Native Son* figure corridors as a metaphor for the novel's mediating role in the representation of modern interiority, she studies how they represent technologies of "corridoricity" such as "ducts, pipes, [and] urban infrastructures" as a way to "refer explicitly to the inscription technologies of the book" and to "thematiz[e] the network of relations between the material printed artifact and its fictionality."[71] Marshall's attention to the book is unusual: while modernist studies has benefited from a wealth of critical attention to print culture, the book has tended to be eclipsed by newspapers, little magazines, and other periodicals. The major exception is Lurz's *The Death of the Book,* which argues for the importance of the book as an object to the modernist novel. As modernist studies begins to contend with the embeddedness of novels in books, I want to suggest that metamedial novels' emphasis on this embeddedness resulted from a merger of modernist writers' preoccupation with form and their awareness of the shift from books to modern information media.

The modernist period was a time when visual artists turned their attention to their media and when poets felt, to quote William Carlos

Williams, that "all the world was going crazy about typographical form."[72] We should add the book to the list of media on which modernist writers focused their formal investigations. Literary experimentation with the visual and tactile dimensions of media during the modernist era tends to be discussed with reference to poetry—the disorienting typographical layouts of the Vorticists, for instance. But novelists, too, played with the printed page. *Ulysses* disrupts conventional textual arrangement with devices such as question-and-answer sets and mock newspaper headlines. William Faulkner attempted to publish *The Sound and the Fury* using several colors of ink to mark narrative temporalities. Photographs became more commonplace; we see them in novels such as André Breton's *Nadja* and Woolf's *Orlando*. Novelists were also more involved in the production of their novels as physical artifacts. Like the artists' books that also commenced as a practice in the early twentieth century, these novels are "self-referential and self-aware objects" that "interrogate the codex, calling into question how books communicate and how we read, using every aspect of their structure, form, and content to make meaning."[73] This novelistic experimentation was not limited to exploring form for its own sake; it also reimagined and revised the relationship between the novel's representative abilities and its medium. As writers reflected on the novel's place in a society dominated by information, their works showed the novel's representation of information culture to be predicated on the book's mediation of information.

FROM THE MODERNIST ERA TO THE TWENTY-FIRST CENTURY: INFORMATION AS THE FORMATION OF DATA

I put the modernist and contemporary periods into dialogue to draw out the rich connections between them. I also do so in order to think broadly about the comparative work of media history. To establish my reading of the early twentieth and twenty-first centuries as parallel moments of media transition, the next four chapters alternate between these eras, with the final chapter serving as a synthesis. A productive alchemy arises from the juxtaposition. This parallax reading brings to light how modernist ideas about literary form chime with ideas about form from information culture, and it clarifies how the novel's archival project has been reimagined through the ways in which books and other information media have been theorized in critical and popular accounts. When we view the contemporary moment as continuing the legacy of modernism's engagement with information, mediation, and form, we better understand our own information culture. *Out of Print* models how we may take into account the way in which one era's media "anticipates" another's (the word is frequently used to locate similarities), without either reinstating a teleological view of history or simply describing the two as uncannily alike without delineating any trajectory between them. It is not quite that modernist information culture gave rise directly to today's information culture, nor is it quite as fuzzy as the statement that the one merely anticipated the other. To identify continuities as well as divergences between these periods is to formulate a more robust theorization of novelists' responses to the scale of information, tracking the persistence of forms across time.

The chapters that follow consider several related but distinct aspects of form central to information, the novel, and the book: form as organization; form as embodiment; and form's role in innovation, preservation, and obsolescence. Taken together, these chapters develop a concept I call the formation of data. As I have stated, I define *information* as data made meaningful by the ways in which it is contextualized and situated. Information is constituted by the forms it takes. The formation of data encompasses how data are organized when they are collected and stored, how they are embodied in media, and how these ways of giving form to data create meaning. Metamedial novels highlight all these facets of the formation of data.

The first two chapters focus on information's arrangement— its organizational form. Chapter 1 describes information shock, modernism's version of information overload, a condition produced by textual proliferation too vast and disorganized to manage. I argue that modernist novelists contributed to debates about information shock by representing the cognitive experiences it created and merging an informational aesthetic with a renewed attention to the form of the book. Focusing on *Ulysses,* I argue that Joyce highlights the novel's textual materiality in order to critique mediation, which I show to be central to the emerging field of information management. I take Paul Otlet's Mundaneum archive as a paradigmatic case study, documenting how the systematic management of information, as it moved away from the book to meet new economies of information scale, entailed methods of organization, processing, and navigation whose details could be hidden from lay users. Against such mediation, *Ulysses* demonstrates how the metamedial novel could reestablish form's centrality to information.

Chapter 2 moves from early twentieth-century information management's use of classification systems to the algorithmic analysis and search systems that dominate mediation in the era of Big Data. In my reading, contemporary novelists' response has been not to represent this scale in and of itself but to explore how data are made meaningful through the limits that contain them. I show how, in texts including Danielewski's *Only Revolutions* and Matthew McIntosh's *theMystery.doc,* narrative, book, and page all become systems that impose a comprehensible order on information. These works challenge conventional definitions of the literary by producing literary text through the distillation of large quantities of information. The value of the book as a novelistic medium, these texts suggest, is its ability to make conceptual and spatial limits visible as frameworks, as the novel itself becomes a meaningful framework for contextualizing information.

The second set of chapters examines the implications of information's embodiment in media—its material form. Chapter 3 argues that the book's tactility and consequent ability to preserve the physical traces of readers took on a new significance during the modernist period. This emphasis on what I call *haptic storage* occurred in response to new regimes of personal information collection that sought to quantify people via data and to the rise of microform as a competitor to the print book for information storage. In this pivotal moment, the use of microform reading interfaces encouraged the rhetorical stance that medial embodiment was inconsequential to information, a principle epitomized by Bob Brown's microfilm-inspired Readies project. I read Woolf's *Orlando* as a critique of this disregard of embodiment and material form. *Orlando* problematizes the idea that any quantity of information could substitute for the presence of a living subject. Instead, it suggests that Woolf's project of capturing the life of her estranged lover becomes realized only when narrative descriptions and the accumulation of biographical detail are combined with the haptic "information" created when readers interact with books.

Chapter 4 analyzes how novelists have used the book's hypermediacy and affordance of haptic storage to challenge the posthumanist principle that a sufficiently large amount of digital data can represent or replace a living subject. I examine this principle in the context of several digital immortality projects, arguing that the shift from print books to digital media has influenced the metaphors and ideologies of digital immortality, in addition to making this research possible. In response to the hope that people could live forever as informational representations, transcending their mortal human bodies, I study several works of contemporary literature that dramatize the potential horror of such existence. I analyze two novels and a memoir-elegy— Steven Hall's *The Raw Shark Texts*, Foer's *Extremely Loud and Incredibly Close*, and Anne Carson's *Nox*, respectively—arguing that they demonstrate how an emphasis on the book's materiality can be useful to the work of mourning. My analysis of these works clarifies how books and digital media have inspired distinct rhetorical positions regarding representation and death because of how they mediate information.

Chapter 5 synthesizes the preceding chapters, turning from the early twentieth and twenty-first centuries to the future. I examine the close links between discourse on the death of the novel and discourse on the death of the book, particularly as both have been influenced by media transition and issues of platform stability. I argue that the book, in its fragility and its persistence, has served as a model for thinking through futurity. I read the novel as a literary form especially suited to theorizing media transition, and I contrast the celebration of innovation in the technology industry with skepticism about innovation with literary form. The chapter analyzes three visions of how the relationships among print books, digital media, and literary reading may manifest in the future: Katie Paterson's Future Library Project, Stephen King's Kindle novella *UR,* and Sebastian Schmieg and Silvio Lorusso's artist's book *56 Broken Kindle Screens.* I conclude by discussing the ongoing relevance of modernist literature and modernist studies for thinking about media, temporality, and futurity.

The novel has been transformed over the last century, and this transformation has intensified with the growing scale of information. Novelists have explored the porous boundaries between what constitutes literary narrative and what constitutes information, and they have leveraged the book's hypermediacy to incorporate the aesthetics of information and critique the assumptions that underlie mediation. Novelists' experiments with the book and literary form reassert form's importance to information, even if the dominant conditions of contemporary information culture work against a conscious awareness of, or interaction with, these forms. To construct a medium-specific approach to the novel is thus not only to consider how the print book has impacted the novel but also to consider how the novel circulates within, and interacts with, its media ecology. The print novel is not a nostalgic form. It continues to shape popular understandings of information even as it adapts to engage with new media, new practices of mediating information, and new ideas about the book.

Information Shock

Systematic Management and the Modernist Novel

The domain of literature . . . in both its modern and postmodern forms has been forced to respond in deep ways to the social condition of information overload.
—John Guillory, "The Memo and Modernity"

This may, or may not, be literature. It is certainly good cataloguing.
—Shane Leslie, 1922 review of Ulysses

TOO MUCH TO READ

In modernism's most famous scene of vexed reading, the Londoners of Virginia Woolf's *Mrs. Dalloway* struggle to decipher an advertisement written in the sky:

> Dropping dead down the aeroplane soared straight up, curved in a loop, . . . and . . . out fluttered behind it a thick ruffled bar of white smoke which curled and wreathed upon the sky in letters. But what letters? A C was it? an E, then an L? Only for a moment did they lie still; then they moved and melted and were rubbed out up in the sky, and the aeroplane . . . began writing a K, an E, a Y perhaps?
>
> "Glaxo," said Mrs. Coates in a strained, awe-stricken voice. . . .
>
> "Kreemo," murmured Mrs. Bletchley, like a sleep-walker. . . .

33

"That's an E," said Mrs. Bletchley. . . .

"It's toffee," murmured Mr. Bowley. . . .

It had gone; it was behind the clouds. . . . Then suddenly, as a train comes out of a tunnel, the aeroplane rushed out of the clouds again, . . . and it soared up and wrote one letter after another—but what word was it writing?[1]

What word do these ephemeral letters spell out? The crowd fails to reach a consensus, disagreeing even on individual characters. The effort of reading arrests the spectators physically and psychologically: one struggling reader is "awe-stricken" while another stares "like a sleep-walker." They are suffering from that most modern of neurological afflictions: shock. The scene has become well-trodden ground in modernist scholarship, in part because many of shock's classic catalysts are present: a city, a crowd, and an airplane (itself evoking technology, speed, and the First World War).[2] Yet Woolf's airplane is more than modern technology or repurposed weaponry. It is an inscription machine. If the skywriting episode of *Mrs. Dalloway* speaks to the modernist novel's interests in themes such as consumer culture, the impact of the Great War, and the elusory nature of signification itself, it also demonstrates the cognitive burden of navigating the diverse and ubiquitous texts that circulated through the modern city.

This chapter examines a condition I call *information shock*: the state of being overwhelmed by the proliferation of texts and textual media. Information shock falls within the spectrum of early twentieth-century shock. Shock could manifest physically, mentally, or both; it could stem from the visceral traumas of trench warfare or the overstimulation of a crowded tram. The condition I am describing takes *information* literally (in both senses of literally): not as any stimuli that must be processed at the neurological level, but as textually inscribed data. An explosion of textual information was straining mental life in the modern metropolis. *"There is too much to read,"* proclaimed a 1920 advertisement in the New York periodical the *Weekly Review*. "You *must* pick and choose with deliberate purpose, or you must flounder."[3] Print publications had reached record numbers, and they permeated public spaces to an unprecedented degree. "Fifteen minutes on a trolley car without something to read has become a horror," wrote Henry Seidel Canby in 1915.[4] But having too much to read had become its own horror. Texts had increased beyond the scale of any reader's control. Although anxieties regarding information overload have a long history, the novelty of information shock during this period stemmed from the sense that information had become unmoored from the book and that the book, as a result, was no longer capable of containing or managing information.

Scholarship in modernist studies has increasingly recognized that the period's information and media cultures, like its print cultures, were integral to modernist literature. This chapter adds the story of information shock—its causes, its social impacts, its influence on the novel, and its implications for the theorization of the book during the modernist era—to this critical work. Analyzing novelistic representations of information shock in relation to media history and the history of information science, I examine how the modernist period's information culture developed in the United Kingdom, Ireland, and the United States. I demonstrate that information proliferation and the systematic methods introduced to mitigate it were major issues well before the Second World War, the event often viewed as the origin point of contemporary information science.[5] Most crucially, my analysis documents how the ruling ideology of modern information culture—that information proliferation could be managed through mediating systems that treated information's form as incidental to its content—was inseparable from the marginalization of the print book. I examine these dynamics through an extended case study: Paul Otlet and Henri La Fontaine's Mundaneum archive, a project whose lofty goal of collecting all of the world's information had far-reaching effects on the development of information science. The Mundaneum instituted a structured, modular, and mediated system for organizing information that typifies how early information professionals moved away from the book to meet the demands of information scale.

In the information culture within which literary modernism emerged, books were deemed inadequate for the task of information management. This context is essential for understanding why and how the novel's archival project changed during the first decades of the twentieth century. As I argued in the introduction, it was during these decades that this project shifted from a realist accounting of the world to a detailed examination of the ways in which novels operate as information media, embedding data into their narratives and drawing on the discursive and formal conventions of information media. This chapter develops the argument further, claiming that modernist writers responded to the conditions I have been describing by refashioning the novel into a medium capable of incorporating challenging quantities of information within the space of the print book. These writers explored how the novel's embodiment in the book could be used to emphasize the inextricability of information's form from its meaning. This strategy contrasts with the management of information via new media and new methods that positioned interfaces between data and users, obscuring this form.

Modernist writers from James Joyce and Virginia Woolf to John Dos Passos and Gertrude Stein were deeply interested in questions of information scale. How might it be represented within a narrative? How might a novel simulate the cognitive burden of information shock for its readers? How could the formal configurations of managed and mediated information provide an aesthetic model for the novel? I focus in particular on *Ulysses,* a text that is at once modernism's most notoriously informational novel and a highly atypical work that raises questions about the generalizability of metamedial novels' archival strategies. Juxtaposing Joyce's famous encyclopedism with Otlet and La Fontaine's grand ambition to archive the entire world, I read *Ulysses* as an insistently bookish work that foregrounds the affordances of the print book in order to stage a critique of totalizing, mediated information systems such as the Mundaneum. As Joyce's contemporaries registered information shock as an inescapable condition of modern urban life, *Ulysses* imagined how one might live with large-scale information without either perfectly managing it or succumbing hopelessly to information shock. I conclude this chapter by considering the issue of exemplarity in the theorization of the novel, arguing that genre, like medium, may best be studied through idiosyncratic examples that disrupt formal conventions.

TEXTUAL EXCESS AND MODERNISM'S MEDIA CULTURE

Information shock was a media problem. First, and most obviously, there was too much reading material. Second, information was distributed across a variety of textual media. Finally, these media circulated ubiquitously in a disorganized and unmanaged fashion. In combination, these factors supported the perception that information had become not only abundant but overwhelming, impossible to contain or control. Consequently, readers sought a means of navigating and managing modernity's streams of information. All of these elements influenced how novelists responded to the information culture of their time.

The first issue was one of information abundance. The quantity of available texts daunted would-be readers. With information shock, scale is a relative measure: *too much* reading matter for any reader to consume. As one hyperbolically exasperated writer described the situation in the *Irish Independent*:

The world is overrun with books. Since Gutenberg invented printing . . . over thirteen million books have been published. And every year Great Britain alone adds . . . seven or eight thousand new books. . . . I don't admire, on moral grounds, the Saracens of old who made a bonfire of the Alexandrian

Library, but I am all for the burning business. . . . If the number of books was reduced, by burning or other means, to a manageable proportion—well, it would be a great relief to readers who, when they go into a modern book-shop are assailed from every side with books that clamour to be read.[6]

Like the *Weekly Review*'s complaint of "too much to read," this writer's frustration stemmed from the profusion of published literature. Of course, the nineteenth century also witnessed a sharp increase in the volume of printed matter, and we can trace the complaint of information overload to antiquity. Ann Blair describes this as "the overabundance of books"—the state of having more books than anyone could reasonably hope to read—and quotes Seneca's assertion that "the abundance of books is distraction" as one example from the ancient world.[7] In other words, twentieth-century anxieties about the scale of publishing were newsworthy but not new: "the announcement that 12,799 books were published during 1926 has elicited a cry against over production, but such elements have been common since the days of Solomon, and perhaps earlier still."[8] Information overload has taken a number of forms over its history, configured differently in each episteme.

The distinguishing factor for the modernist period is that the volume and variety of publications made this perception a well-established, everyday reality rather than the specialized lament of scholars. The publishing culture of the early twentieth century solidified the technological and social changes that had been increasing print production since the mid-nineteenth century. Improvements in publishing technologies, including the invention of Linotype and the development of wood-based paper, drove large increases in book publishing. Meanwhile, education reform increased demand for reading material.[9] As I noted in the introduction, book production rates increased dramatically in both the United Kingdom and the United States between the mid-nineteenth century and the first decade of the twentieth century. While the rates of books published in Ireland were considerably lower, and while the Irish were less likely to purchase books than were their English and American counterparts, the Irish reading public did consume books imported from Britain, and influential Irish publishers such as the Dun Emer Press and the Talbot Press were founded during the first two decades of the twentieth century.[10] These changes were felt widely. The question "are too many books written and published?" even became the subject of a radio debate, broadcast by the BBC in 1927 and presented by Virginia and Leonard Woolf. Virginia answered in the negative.

The Woolfs' debate, like much of the wider debate about book production, centered on the quality of literature being written and the public's ability to choose worthy reading material.[11] These were anxieties about class, education, and social gatekeeping. Aldous Huxley opined in 1934 along these lines, saying that "for every page of print and pictures published a century ago, twenty or perhaps even a hundred pages are published today. But for every man of talent then living, there are [now] only two men of talent," with the result that "the proportion of trash in the total artistic output is greater now than at any other period."[12] The conversation was often couched in the language of shock. Leonard Woolf described book publishing as dehumanizing and overwhelming: once an art, it had become "in the machine age" a mechanized dystopia, "a large scale industry, with books dictated to shorthand writers, typed up on typewriters, and set up on monotype machines."[13] One imagines a *Modern Times*-esque factory, with dazed workers churning out bestsellers. That this resulted in "a flooding of the market" left readers in a similarly shocked state.[14] Thus, publications such as the *Times Literary Supplement* promoted themselves by claiming that they would solve the "very rea[l] problem . . . of choosing books from among the *bewildering mass* that *ceaselessly pours from the press*."[15] The fact that this advertisement included a typographical error ("very rea problem," omitting the "l") itself links urgency and distraction with the proliferation of print media.

Guides such as the *Times Literary Supplement* were another indication that the reading public viewed the abundance of reading matter as precipitating a cognitive crisis. Reading strategies became popular ways to cope. One strategy was to filter what one read. Many periodicals offered columns summarizing recent literature in order to save readers the time and attention required to navigate the field themselves. Patrick Collier has shown that *John O'London's Weekly* "offered shortcuts to becoming 'well-read'" such as "reviews, which were heavy on summary and quotation" and "*Tit-bits*-style paragraphs . . . , which presented paragraphs-long quotations from recent books with minimal content."[16] In the United States, *Reader's Digest* was founded in 1922; it offered readers collections of condensed works for quicker perusal. By the mid-1920s, book-of-the-month clubs had been established in the United States and the United Kingdom. These, too, allowed their patrons to relinquish the work of choosing what to read. Even children, it seemed, needed guides: *A Bibliography of Children's Reading* in 1908 cautioned that, because "the bookish boy has too much to read," he may be in danger of haphazardly "hurry[ing] through one book to another."[17] Children and adults alike were urged to implement a combination of guides and reading schedules. In *Ulysses,* Stephen Dedalus mocks the reading schedule he adopted when a boy: "Reading two pages apiece of seven books every night, eh? I was young."[18] The promotional rhetoric of readers' guides assumed that the cognitive burden of information shock was real.

Such guides functioned as interfaces by which readers could navigate the flood of new publications indirectly, ignoring those not deemed worthy. Their existence reinforced the perception that, without help, a reader would be unable to cope with the shock of having too much to read. Unaided, a reader might be so overwhelmed as to give up on reading entirely or fall into the trap of reading extensively rather than intensively. Better strategic reading or no reading, some guides urged, than bad reading. The *Weekly Review* advised its subscribers to stick to a strict "schedule" of reading: "Not more than two daily papers.

Not more than two weekly journals. Not more than four monthly magazines." Without this intervention, advertisements and articles warned, readers would lapse into "bad" reading: "We read too much to read intelligently. We are bad readers."[19] Such claims assume that the sheer quantity of textual matter in the world was impossible to navigate without experts, who could mediate between the "bewildering mass" of publications and the apparently bewildered masses of the reading public.

The rhetorical position that the volume of published reading material had become mentally taxing was pervasive. This was a problem of information rather than media, of what and how much is read rather than the forms that content takes. Taken on its own, the fact of textual abundance does not demarcate modernism's information culture from those of preceding eras, although modernism's abundance was publicly acknowledged on a wider scale. The significant change was that this lament of too much to read—of information shock—became as much an issue of medium as of content. The category of medium came to the forefront during this period as print faced competition from new storage technologies. One of the distinguishing factors of modernism's media culture was variety. With the introduction of devices that could store and reproduce visual and audio data, information was no longer synonymous with the written record. As Friedrich Kittler has argued, new nontextual media such as the gramophone and the cinema had a huge impact on the literary imagination: they challenged the novel, offering competing modes of narrative and representation.[20] But so did textual media, both in the emergence of new varieties and in the changing uses and circulation patterns of established ones. As writers grappled with the novel's evolving role as an archive, they had to contend not only with the growing importance of information storage as a cultural phenomenon but also with its changing nature.

Newspapers and magazines had altered the landscape of textual media. As Ann Ardis and Patrick Collier describe, "the turn of the twentieth century saw a sea change in the world of Anglo-American book, newspaper, and periodical publishing."[21] A major part of this change was "the expansion of a highly differentiated periodical culture," including newspapers, magazines, and little magazines.[22] Indeed, the *Weekly Review*'s advice that readers needed to strategically pick what to read attributed the cause of overload not to books but to "the teeming periodicals of to-day."[23] The first British halfpenny daily newspaper began circulating in 1896, the first mass-circulated tabloid-format newspaper in 1903; and the 1920s and 1930s witnessed a substantial increase in the circulation of daily newspapers.[24] "The Englishman is overwhelmed every morning with a white spray of facts," wrote Ford Madox Ford of newspapers, a description capturing the bodily sensation of information shock.[25] The situation was similar in the United States, where the number of cities with a daily newspaper more than trebled between 1880 and 1910 and circulation rates climbed from roughly 22 million in 1910 to nearly 40 million during the next two decades.[26] In Ireland, too, where the demand for newspapers and periodicals exceeded that for books, there was "a great expansion" between the turn of the nineteenth and twentieth centuries, "with the spread of newspaper publishing to almost every Irish town of any size."[27] In the first decade of the twentieth century, a British Census of Production estimated that books made up less than 15 percent of all printed items.[28] Walter Benjamin names newspapers as a contributor to the shock experience (*Chockerlebnis*) of the modern city: he cites both urban traffic and "the advertising pages of a newspaper" as examples of optical shock.[29]

Media also diversified in other ways. The skywriting in *Mrs. Dalloway* is an extreme example, but it speaks to how thoroughly textual media permeated the modern city. When E. M. Forster imagined "collect[ing] together all the printed matter of the world into a single heap," he pictured not only books and newspapers but also public writing: "advertisements" and "street notices." Tellingly, his list ends with a word indicating the multiform nature of textual media and the difficulty of listing them comprehensively: "everything."[30] To walk through a city was to be subjected to a barrage of textual information in any number of forms: booksellers' stalls, newspaper boys, vendors selling a range of periodicals, and advertising posters plastered to buildings and trams. Signage had become, in Paul Stephens's words, "the ambient background to everyday life."[31] Modernism's writers were born into a world where "the most common reading experience . . . would most likely be the advertising poster, all the tickets, handbills and forms generated by an industrial society, and the daily or weekly paper."[32] These conditions, begun in the nineteenth century, had intensified by the time of modernism, spurred by urbanization and the burgeoning advertising industry.

In private and professional spaces, readers also encountered a large volume and variety of textual media. In the office, businessmen and clerks read new formats and genres such as memos, telegraphs, tickertape, and microfilm, as secretaries produced typescript and mimeographs. In the cinema, pleasure seekers read intertitles. As printed matter proliferated at heightened rates and in diverse forms, it also circulated at higher volumes through public and private spaces. Information shock resulted not only from the abundance of textual materials but also from the sense that their variety and ubiquity made them difficult or impossible to manage. This diversification of texts, proliferating across different social contexts, media, and genres, is what makes the situation I have been describing an issue of information rather than only of text or media. As the information historian Ronald E. Day notes, it was during the modernist period that early information scientists such as Paul Otlet began to view "reading [as] information transfer."[33]

That information shock was a recognized and widespread problem is evident throughout modernist literature. Again and again, characters struggle to read, and this struggle is tied to a consistent representation of the modern world as a space of disorganized, uncontrollable, informational texts that cause cognitive overload. Characters may struggle to read for other reasons; they may be distracted by personal concerns, for instance. Frequently, however, the reason is both mundane and modern: the proliferation, variety, and ubiquity of texts. While the urban texts of *Mrs. Dalloway* reach into the sky, in *Jacob's Room* they descend into subterranean realms, in the "large letters upon enamel plates" lining Underground stations.[34] It is as if Filippo Marinetti's visual depiction of chaotic texts jostling across a page amid cars and airplanes is less avant-garde experimentation and more realist representation of cities as textual spaces.[35]

A media historical approach to modernism reveals textual media as one of the many stimuli that city dwellers had to manage, habituate, or ignore. In John Dos Passos's *Manhattan Transfer*, for example, George Baldwin frowns

> at the gold lettering through the ground-glass door.
> NIWDLAB EGROEG
> WAL-TA-YENROTTA
> Niwdlab, Welsh. He jumped to his feet. I've read
> that damn sign backwards every day for three months.
> I'm going crazy.[36]

Dos Passos prints the words backward, mimicking the effect of reading a sign through a glass door and forcing the reader (like Baldwin) to decode it. The frustration of the scene is palpable, as Baldwin realizes that the text of his own name has become a meaningless signifier to him. His misreading is a consequence of the ever-present textuality of urban life: as soon as he leaves his office building, he is confronted by "the headline on a pink extra."[37] The scene will be repeated again and again as characters read the newspapers, signs, and advertisements that circulate in New York City.

In Joyce's Dublin, similarly, characters are so constantly bombarded by textual media——fliers, newspapers, postcards, advertisements, and letters, to say nothing of books——that they tend to skim or read at

44

random. When Leopold Bloom is handed "a throwaway," he quickly skims its contents:

> Are you saved? All are washed in the blood of the lamb.
> God wants blood victim. Birth, hymen, martyr, war,
> foundation of a building, sacrifice, kidney burntoffer-
> ing, druids' altars. Elijah is coming. Dr John Alexander
> Dowie restorer of the church in Zion is coming.
>   Is coming! Is coming!! Is coming!!!
>   All heartily welcome. (8.10–15)

*Information
Shock*

Like Dos Passos, Joyce uses typography—here, the increasing exclamation marks (and, in the first edition of *Ulysses*, italicization)—to mimic a diegetic printed text within the space of the print book. (I will return to the significance of such metamedial moves later in the chapter.) Bloom glances over the flier; he does not have time to read it thoroughly, for he will soon have other texts to read. (A few lines later, he crosses the path of men wearing sandwich-board advertisements.) This nonlinear approach to reading serves him for books, too: "the *sortes* technique [of choosing a page at random] is actually the most common way Bloom reads."[38] Brief bursts of attention are apropos for an advertisement canvasser like Bloom. As one piece in the advertising trade journal *Printer's Ink* put it in 1908, because "we live in a busy time, with too much to read, the brevity [of an advertisement] that invites, and demands only moderate and quick attention, may talk to more readers than an announcement of considerable length."[39]

Neither skimming nor aleatory selection were new, but *Ulysses* depicts these practices as necessary rather than optional. We might recognize the cognitive ancestor of the Internet era's distracted reading in such scenes. Whereas contemporary readers may develop hyper attention as a strategy for coping with digital culture's distracted reading, the conditions in the early twentieth century seemed not so

much something to be adapted to as something to be endured.[40] This
aleatory skimming was the "bad reading" that readers' guides warned
against, but, in the context of city streets, there was no way to filter
irrelevant content so that one could read only worthy texts with proper

attention. These opposing approaches to information shock—reading
selectively and reading at random—both assume that readers cannot
confront the scale of information in its entirety and so must find a
way to navigate portions of it.

Additionally, information frequently came in the form of isolated,
decontextualized fragments, atomized sections of text such as headlines
or advertising slogans. These texts were encountered piecemeal, without an understanding of their relation to the entire documents they
derived from, to other texts, or to larger organizational frameworks—
without an understanding, in other words, of the contexts that would
make them meaningful. In this respect, literary material circulates
like informational material: Leopold Bloom and Clarissa Dalloway
are as likely to recall a line from *Sweets of Sin* or *Cymbeline* as they are
to reflect on a line from an advertisement they pass. Literary texts,
too, become fragmented within the landscape of textual information.
As modernist novels expose these conditions, their representations
of reading investigate how information functions, and breaks down,
at large scale.

Thus, information shock raised questions about the aesthetics of
information: how a literary text might represent the experience of
navigating textual environments within a print book (literary aesthetics) and how information itself entails form (information *as* aesthetics).
Information emerges out of the interplay between embodied form
and organizational form. The texts moving through the modern city
seemed to resist formation into organizing frameworks. In a library,
books are organized according to certain principles; their call numbers and other identifiers map them into a conceptual web, locating
each object in relation to all others. If a library user knows the call
numbers for *Manhattan Transfer* and *Mrs. Dalloway,* this is meaningful

information about how each relates to the other as categories of literature, how each is located in space relative to the other books in the library, and where to find each novel on the shelf. Fliers, advertising posters, newspaper headlines, skywriting, and other unbound and mobile texts have independent trajectories. The majority bear no relation to one another, spatially or semantically. Information shock arose from unstructured scale, information proliferation lacking a clear organizing principle.

FROM BOOK TO FILE: MANAGEMENT, MEDIATION, AND THE MUNDANEUM

I have been describing information shock as it impacted individuals and public spaces. In institutional contexts, too, textual information proliferated. While the woman in the street had no way to order the texts she encountered, her office worker counterpart used systems to organize data and processes to manage the flow of documents. The first decades of the twentieth century solidified a field that had been growing since the middle of the nineteenth century: the systematic management of information. From clerks, secretaries, and office managers to professional librarians, information management was an increasingly significant industry. By 1910, what we would now designate information work had become established as a professional discipline, in response to information proliferation.

The library systems in Britain, Ireland, and the United States, for example, had all expanded considerably by the early twentieth century. The number of library authorities in the United Kingdom grew from only twenty-seven in 1868 to more than five hundred in 1918.[41] In 1867, there were just over 1,000 public libraries in the United States; the number was nearing 3,000 by 1917.[42] Although the growth of libraries was slower in Ireland, the numbers of library patrons and books borrowed increased across the country, particularly in the first decade of the twentieth century.[43] As library systems increased their holdings and opened new branches, many librarians worried that their collections were out of control. In 1902, Sidney Webb wrote that "the library service of a great city can and surely ought to be something more than a couple of hundred almost accidental heaps of miscellaneous volumes, each maintained and managed in jealous isolation from the rest, and limited in its public utility by the lack of communication between the heaps."[44] Webb's description characterizes information management in formal terms: "heap" suggests a form at once identifiable *as* a form (as Sianne Ngai points out, a heap is a recognizable figure, a grouping that contains) and formless (a shapeless mass, the accumulation of fragments without internal structure).[45] Two and a half decades later, at a talk for the Association of Special Libraries and Information Bureaus, A. D. Lindsay stated that the "terrific accumulation of the materials and instruments of knowledge" was causing "mental indigestion."[46]

Businesses also needed to manage information proliferation. As railroads and long-distance communications technologies such as the telephone and telegraph allowed larger business enterprises to flourish, businesses generated larger volumes of documents to oversee operations: "The informal and primarily oral mode of [business] interaction gave way to a complex and extensive formal communication system depending heavily on written documents."[47] These documents included new formats such as memos and other standardized forms. The growth in documents helped to coordinate operations but created a feedback loop in which businesses could be overwhelmed by the spike in paperwork:

> Less than two generations ago the businessman . . . who had sufficient correspondence to require the services of an amanuensis was the exception rather than the rule. The number of papers handled was so limited that the filing question was a more or less negligible one. . . . Since then the growth of the country, changes in conditions and new inventions have made it possible to transact business in hours and even in

minutes that formerly required days, weeks, or months, and
with this growth has come an enormous increase in the volume of papers to be handled.[48]

Enough paperwork was generated that businesses began to develop *Information*
policies for the destruction of files, and waste-paper traders became a *Shock*
new industry.[49] In response, institutions developed methods of sorting,
classifying, and processing the documents they accumulated.

Critical accounts of literary modernism typically, and rightly,
emphasize the movement's fascination with chaos and fragmentation.
This aesthetic is especially noteworthy when contrasted with information professionals' drive for order. An institutional emphasis on systems
arose as businesses, libraries, and government agencies dealt with a
flood of material that threatened to exceed the capabilities of existing
management practices. Yet the dominant institutional response was
optimism, driven by the pragmatic belief that new media and new management processes could control the situation. "Dam the overflow" of
office paperwork, an advertisement for Vertex Vertical-Expanding File
Pockets urged, featuring an illustration of a mass of documents walled
into place with filing folders.[50] The belief that guided the systematic
management of information was that no amount of information was
too much, given a system capable of breaking all information down
into its constituent parts, classifying each part, and arranging them
within an ordered structure. With the right system, many affirmed,
one could organize all of the information in the world.

The model par excellence for such a system was the Mundaneum, a
Belgian archival project at once hyperbolically grandiose in its aims and
wide-reaching in its influence on the development of twentieth-century
information management. The Mundaneum demonstrates how the use
of modern organizational practices resulted in more mediated experiences of information. It also dramatizes the extent to which books
became peripheral to information management during this period.
Finally, the Mundaneum emblematizes the scope of information work

during the modernist period, providing a counterpoint to novelistic projects of archiving cultures dominated by new scales of information. At the Brussels World's Fair in 1910, the Belgian visionaries Paul Otlet and Henri La Fontaine introduced the Mundaneum to the world.[51] Their goal was as breathtakingly ambitious as it was straightforward: to collect and organize all the information in the world. They believed the Mundaneum would aid research and international scholarship; it might even lead to world peace.[52] Part archive, part museum, and part bibliographic system, the Mundaneum, they claimed, would comprise "one universal body of documentation," "an encyclopedic survey of human knowledge," and "an enormous intellectual warehouse of books, documents, catalogues and scientific objects." "These collections," they continued, "will tend progressively to constitute a permanent and complete representation of the entire world."[53] Le Corbusier agreed, describing the Mundaneum as portraying "the whole of human history from its origins."[54]

The Mundaneum grew out of fifteen years of work in information management. In 1895, Otlet had established the Universal Bibliographic Repertory (RBU), a catalogue of cards that stored the bibliographic details of scientific publications, organized by his Universal Decimal Classification (UDC) system.[55] He hoped to establish an international index providing bibliographical data for every publication since the invention of the printing press. In less than a decade, the RBU contained 3 million entries, prompting Otlet and La Fontaine to found the International Institute of Bibliography (IIB) to oversee the project—and, eventually, to expand it into the Mundaneum.[56] By the Mundaneum's 1910 debut, its elements included a collection of scientific publications; a Museum of the Press, whose goal was to archive the first and last issue of every newspaper ever published; objects and artifacts forming an International Museum; and a collection of ephemeral graphical works such as posters. The RBU grew to more than 15 million entries by 1930; the collections of documents and objects also grew. Today, despite the fact that more than ninety tons of material were destroyed between 1970 and 1980 (in addition to materials lost or destroyed during the Second World War), the remaining material occupies more than three miles of shelf space.

It is easy to overlook the Mundaneum's indebtedness to, and influence on, the information practices of its era. On the one hand, it is an iteration of the old dream of a vast archive containing all the world's knowledge, a trope found in texts as diverse as Denis Diderot's *Encyclopédie* and Jorge Luis Borges's "Library of Babel." On the other, the Mundaneum prefigures contemporary information culture: to many, Otlet and La Fontaine are the founding fathers of information science, and descriptions of the Mundaneum frequently emphasize its anticipation of the Internet. Thus, the Mundaneum is "the Web that time forgot" or "Steampunk hypertext."[57] Yet the Mundaneum was a product of its time, similar in spirit to other projects that sought to record the world, including H. G. Wells's permanent world encyclopedia and, as the next section examines, encyclopedic novels such as *Ulysses*.[58] Otlet and La Fontaine believed in the possibility of a total archive because they adopted the modern principles of systematic management to information management. Rather than being frustrated or disoriented by the scale of information, they embraced it. Otlet, too, wrote that "there [was] too much to read": "once, one read," whereas "today one refers to, checks through, skims."[59] For him, however, substituting these other approaches for reading was an entirely viable strategy, given the right system of reference.

Otlet's system had three basic principles: strip each text down to its key data, index the metadata (that is, create a card for the RBU), and classify and file both text and card using the UDC. The first step, which he called the monographic principle, provided the foundational logic.[60] His idea, as W. Boyd Rayward explains, was "to 'detach' what the book amalgamates, to reduce all that is complex to its elements and to devote a page [or card] to each."[61] These pieces could then be moved into a new system, searchable via the UDC. This process depended on atomization and the implementation of protocols used to break down, file, and retrieve materials. As John Guillory has argued, the principle of atomization was similarly at work in informational genres such as the form and the memo, which "divid[ed] up the page into fields, . . . offering boxes to fill or check rather than sentences to write."[62] Otlet's atomization process occurred both physically and in terms of content: the card in the RBU became a representation of the document, often condensing its key points, and the document itself might be disassembled, its constituent parts cut and pasted onto a summary page. The Mundaneum rejected the form of the book, opting instead for a system that could process pieces of data individually, "strip[ping] each article or chapter in a book of whatever is a matter of fine language or repetition or padding and . . . collect[ing] separately on cards whatever is new and adds knowledge," "winnowing" the text in order "to conserve the best grain."[63]

This point is worth dwelling on: the Mundaneum operated on the assumption that books were not suited to modern information management. Their static, bound nature—a strong convention, if not an absolute quality—was a poor match for atomization. Books were precisely the wrong scale: not small enough to correspond to the individual atomized datum, not large enough to provide a conceptual system capable of organizing all possible data. While books had discrete sections, such as chapters, these were fixed in their binding; elements could not be filed separately without destroying the book. The length of books made them hard to copy, and their bound nature made them hard to update. Although Otlet frequently returned to the book as a metaphor for his projects, his grand vision was of a *réseau,* a network. By 1934, Otlet had decided that "the book is only a means to an end. Other means exist and as gradually they become more effective than the book, they are substituted for it."[64]

Otlet and his contemporaries turned from the book to the file and the index card. These modular media were better suited to the factory-like processes of systematic management. Otlet understood himself to be among Frederick Winslow Taylor's followers, extending scientific management to the realm of documentation.[65] Files predate the modernist period, but by Otlet's time filing had become a detailed system, supported by specific media, technologies, work practices, and professional training programs. Filing systems facilitated quick retrieval. "Files are not merely resting-places in which old papers can be buried," stated *Filing as a Profession for Women* (1919); "they are—or should be—libraries of useful information, quickly available when demanded."[66] Loose-leaf binders and vertical filing systems (invented in the 1880s and 1890s, respectively) allowed for assembly line–like workflow in offices. At least one publisher of encyclopedias followed this turn from the book to modular media, advertising a 1910 loose-leaf encyclopedia as "a book that never grows old."[67] Index cards were similarly important. The majority of the RBU's entries were stored on three-by-five index cards, a format Otlet was instrumental in standardizing. Although Otlet would consistently look to new, non-paper-based technologies—he advocated for microform, a medium I discuss in chapter 3—index cards were his ideal medium. A new card could be added to the card catalogue with each new publication, and cards could easily be removed, reordered, or otherwise updated. The card system, he wrote, allowed for "continuous interfiling"; "it alone permits the formulation of the catalogue from contributions coming in from everywhere."[68]

The first principle of Otlet and La Fontaine's systematic management, then, was to break down information into its constituent parts and metadata, which could be stored via unbound media. The second was the organizational system: Universal Decimal Classification, first published in installments between 1902 and 1907 and originally based on Melvil Dewey's Decimal Classification. In the UDC, knowledge was divided into ten categories, each of which was subdivided into a further ten, and so on. Any document or item should, in theory, be classifiable according to one of these subdivisions. (The first edition listed more than 30,000.) Otlet's illustrations emphasized the UDC's neatness and order, picturing it as a subdivided circle. These visual metaphors portrayed his system as complete, with each item clearly oriented relative to the others—"knowledge as a limited circle whose precious elements, namely facts, are stable and can easily be identified."[69] Many of Otlet's and La Fontaine's colleagues shared their belief that the UDC would be a complete system, ordering all the world's information. The Third Report of the International Committee of Decimal Classification describes the UDC's American edition as "cover[ing] practically the totality of human knowledge": "Although I do not believe we can say that absolutely nothing has been forgotten, it will be difficult to mention a topic, for which there is no rubric available."[70] This sense of total comprehensiveness stemmed from the idea that the UDC as a whole was static. Entries were updated and new subdivisions added as necessary, and an early approach to faceted classification allowed for the combination of several subdivisions into a single classification number. Yet the overall system—its ten main categories, subdivided and then subdivided further and further—would be stable. The system's flexibility occurred at the level of media rather than organization: individual documents such as files or index cards could be updated, moved, or altered, but the UDC itself would remain stable.

I have chosen the Mundaneum for an extended case study because it both reflected and directly influenced wider trends in early-twentieth-century information management. Otlet and La Fontaine were influenced by Fordism and Taylorism, and the Mundaneum's systematic management is symptomatic of a time when, as JoAnne Yates writes in her seminal study of business management, "*system, efficiency,* and *scientific* became catchwords in the business world and beyond."[71] Otlet's theories of information management, in turn, influenced information specialists in the Anglophone world. In the United States, G. S. Josephson recommended that the Special Libraries Association adopt his monographic principle: "We might even come to the point where special libraries will not even have whole books, but only such parts of many books as it needs."[72] Otlet's documentation movement was more influential still in the United Kingdom, where the British affiliate of the IIB was founded in 1927.

The Mundaneum demonstrates that considerations of form were inextricable from modern information management. Discursively, organizationally, and medially, the Mundaneum was founded on a careful interplay between parts and wholes, set within a system in which individual parts might move flexibly (say, by retrieving an index card) but the collection as a whole would maintain a strict set of formal relations among items based on how they were organized conceptually (via the UDC) and physically (in card catalogues). Yet this system so focused on form also functioned as an interface, making that form invisible to most of the Mundaneum's users. Systems such as Otlet's were part of a shift toward specialized information access: "Replacing the nineteenth century 'reader' . . . , the 'user' appeared, and the user had to know the terminology and syntax of controlled vocabulary lists, classification structures, and other indexing devices in order to retrieve documents."[73] These systems mediated between would-be reader (or user) and text. Long before computational search—and before the invention of Emmanuel Goldberg's Statistical Machine, which could perform automated searches on microfilmed texts as early as 1931—the systems used to manage textual information created a material interface, a series of card catalogues, folders, and other storage apparatuses that users had to navigate using whatever organizational schemas were in place.[74] This interface was also procedural, operating through the routinized implementation of formal, standardized instructions. This aspect of modern information mediation was once again indebted to Taylor's model: as Alan Liu argues, Taylor's use of instruction cards to manage organizational procedures "might . . . be said to be the first economically or socially significant form of programming" because they turned work into a "structured, modular, and algorithmically manageable process."[75]

The skill required to navigate such systems and implement such procedures resulted in a social interface layer, in the new class of information specialists. As one speaker at a Special Libraries Congress in Oxford put it in 1926, "there are far too many books in the world, and the ideal thing is that a man desiring to acquire knowledge in a particular subject should be guided as to the books specifically of value to him."[76] Ireland's Cumann na Leabharlann (Library Association) was founded in 1904, the Special Libraries Association was founded in the United States in 1909, and the United Kingdom's first library school opened a decade later. Even as libraries began to allow readers to browse their stacks, the increased use of complex bibliographic systems, coupled with the institutionalization of reference services in public libraries, created a division between lay users and library professionals for the retrieval of information.

Specialization also became the norm in business, where the market for filing professionals boomed. Institutions met the demands of their expanding collections of documents by increasing staff specialization and adopting systematic management. While it would not have been impossible for a researcher to use the UDC to look up a text in the Mundaneum as a public library user might locate a title by call number, patrons were more likely to ask the Mundaneum's staff to locate their required items. That the new filing systems were too complex for nonspecialists is evident in this passage from a filing manual: "Much error is caused by the custom of letting outsiders touch the files. A clerk from another department may help herself to a letter desired, perhaps in the absence of the file clerk; or she may replace it for similar reasons, and nine times out of ten she puts it in the wrong place. There should be an iron-clad rule that no one but the file executive or file clerks should touch the files."[77] The Mundaneum offered a mail-in information retrieval service, and the staff fielded more than 1,500 requests annually.[78] This interface function explains why the Mundaneum is today frequently referred to as the "Google of Paper."[79]

The Mundaneum epitomizes how form was both central to and occluded from modern information management. The reader in the city gave up trying to understand the entirety of the information circulating around her, for there was no system to make it conceptually legible. In the Mundaneum, such a system was in place, but the user did not need to—and was encouraged not to—understand it. In both cases, the part eclipsed the whole, and there was little common comprehension of how any one piece related to any other. I have used the phrase "information management" in this chapter because it is common terminology. But this situation was more rightly data management because the concept of information entails an understanding of how individual pieces relate to a larger system of meaning. The Mundaneum exemplifies the paradox of modern information management: a system theoretically capable of organizing every conceivable text into a gestalt—a meaningful entirety, a whole with internal semantic form— made it no longer necessary for individuals to understand each text as part of that gestalt.

"GOOD CATALOGUING": INFORMATION ABUNDANCE AND METAMEDIALITY IN JAMES JOYCE'S *ULYSSES*

As we have seen, early twentieth-century information culture was dominated by questions of scale. How should readers cope with information shock? What role should books take in storing and managing large amounts of information? What are the structures that information takes as it circulates or is stored, how are they meaningful, and what are the consequences of systematic management's mediation of these forms? The novel offered a framework for addressing these questions. This was particularly true of the modernist novel, whose tendency to veer into encyclopedic modes linked it directly to considerations of scale and form.

If any of Otlet's contemporaries could rival his archival obsession, it was James Joyce. There is no more concerted novelistic investigation of the problems and possibilities raised by large-scale information, or of how these issues are shaped by books and other media, than Joyce's *Ulysses*. His famous claim that Dublin could be rebuilt based on the perfection of his representation of the city in *Ulysses* has, despite its hyperbole, shaped the novel's reception. Whether or not Joyceans take the claim seriously, many have adopted the author's rhetoric; thus, *Ulysses* is a "fictional Baedeker of Dublin" or "the entire archive of culture."[80] The novel is archival not only in its meticulous documentation of 1904 Dublin but also in the extent to which Joyce consulted informational records to add precise, factual data to it. Given *Ulysses*'s scale and information density, it is unsurprising that the novel has garnered a reputation of being more revered than read. Joyce teasingly described the work as "usylessly unreadable."[81] *Ulysses*'s difficulty stems not only from its wealth of detail and allusion but also from how it thematizes the organization of information. The novel is both a literary archive of Dublin and a meta-archive, examining how information systems—including narratives and books—store and organize information. If *Ulysses* causes information shock for some readers, it also models a novelistic mode of information management.

Ulysses is a singularly singular work. It is not obvious that one could read Joyce's novel as exemplifying anything about modernist novels more generally. Nor is the designation "novel" self-evident. Franco Moretti, for example, classifies *Ulysses* as a "modern epic" rather than a novel. For him, modern epics (his examples include poetry and opera as well as prose works) may have "encyclopaedic ambition," but they are distinct from novels.[82] Yet to discard the category of the novel when discussing *Ulysses* is to overlook three crucial, interrelated points. First, the theory of the novel has been constituted through an examination of outlier texts such as *Ulysses*: the novel is a genre formed through its exceptions. (I will return to this point at the conclusion of this chapter.)

Second, in its very qualities that make it atypical—its information density, its stylistic experimentation—*Ulysses* exemplifies a wider interest among modernist writers in the novel's relationship to information culture. Novels have been concerned with information since their inception. During the modernist period, this interest focused on the issue of information proliferation. This is one reason why so many modernist texts operate in an encyclopedic mode.[83] We cannot disentangle formal scale in modernist literary works from larger conversations about the scale and management of information. For example, Paul Stephens reads Gertrude Stein's information-dense poetics as a method of creating altered states of attention in the reader, a strategy he attributes to the fact that Stein "considered information excess to be a serious problem" and "was both attracted to, and repelled by, American industrial production and the increasing organization and standardization it required."[84] Consider how she employs the language of classification and systematization at the beginning of *Tender Buttons,* her catalogue of domesticity: "a *kind* in glass and a cousin, a spectacle and nothing strange a single hurt color and an *arrangement* in a *system* to pointing. All this and not ordinary, not *unordered* in not *resembling.* The *difference* is spreading."[85] A novel such as *Mrs. Dalloway* may lack *Ulysses*'s extreme detailism and *Tender Buttons*'s dense semantics, but it, too, explores how narratives may be formed out of assemblages of individual impressions, observations of urban space, intertextual allusions, and other information-rich cultural fragments.

Finally, the way in which novels from this period mediate cultural information is grounded in an awareness of how the genre of the novel and the medium of the book interoperate. Modernist authors frequently published chapters in venues such as little magazines, but books remained the genre's default final form.[86] Additionally, with the rise of small presses such as the Hogarth, Hours, and Cuala presses, writers had more say regarding the form their books took as material objects. This tied novelistic explorations of information scale to the very medium that modern systematic management found lacking. This was certainly true in the case of *Ulysses,* whose design was carefully managed by Joyce. When I argue that *Ulysses*'s project of information management is grounded in a bookish aesthetics—and that *Ulysses* mobilizes the reader's awareness of the way its physical form bears on its organization of information—I thus locate it as part of a novelistic tradition of emphasizing the genre's long-standing association with the print book.

We have already seen that modernist novels represented the cognitive burden of information shock within their narratives. *Ulysses* demonstrates how a novel could also address information scale through metamedial emphasis on the book. Joycean criticism has tended to focus on Joyce's formal innovations in terms of linguistic style; *Ulysses's* metamedial dimensions are comparatively undertheorized. One exception is John Lurz's work on *Ulysses* and *Finnegans Wake,* which makes the case that Joyce systematically draws the reader's attention to the text on the page. For instance, Lurz notes that Bloom's first word in the novel is "O," a word-letter that uses "the book's type . . . [to offer] the reader an icon that depicts the shape [Bloom's] mouth must make to speak aloud."[87] Lurz reads this sustained emphasis on the book as a means of demonstrating how the processes of reading implicate the reader's body and attention, invoking the passage of time and the inevitability of death. *Ulysses's* metamediality also foregrounds the book's role in the mediation of information.

In its self-reflexive play with its medium, *Ulysses* explores how novels, like books, accumulate and organize information. Mapped onto the *Odyssey, Ulysses* invites the reader to order and classify, matching Joyce's narrative to Homer's. To quote T. S. Eliot, Joyce uses myth as "a way of controlling, of ordering."[88] The schemas Joyce composed for Stuart Gilbert and Carlo Linati add additional layers of order: assigning each episode in *Ulysses* a color, symbol, organ, art, and technic, these schemas imply that *Ulysses* is not a random assortment of details. Visually, too, many episodes are structured by distinctive typographic arrangements. In the first edition of *Ulysses,* the "Aeolus" episode, which takes place in a newsroom, is broken into sections, each indicated by a headline, printed in all capitals, centered, and set in a different and slightly larger font than the rest of the text. In "Ithaca," the text clusters on the page in sets of questions and answers, set off by indentations and skipped lines. Each section break in "Wandering Rocks" is marked with three asterisks arranged as a triangle, as if visually representing the dangerous rocks moving across the page.

The hallucinogenic narrative of "Circe" finds its typographic representation in the dizzying page layout, which juxtaposes standard font with capitalized headings and many italicized quotations in smaller font. Even "Penelope," whose 20,000-plus words are grouped into only eight sentences and are almost completely unpunctuated, highlights the importance of visual organization precisely by departing from the norms of the printed page. Foregrounding its textual arrangements, *Ulysses* has as much common in with early-twentieth-century informational documents as it does with premodernist novels.[89]

As the episodes' different typographical styles call attention to the printed page, cueing the reader to notice the novel's material embodiment, the narrative connects print books with information storage. Books are ubiquitous within *Ulysses*. Bloom and Stephen each visit the bookseller's cart; Bloom brings Molly books as love tokens; an entire chapter is set inside Ireland's National Library. I will stop short of trying to catalogue all of the books in *Ulysses;* the word *book* appears in some form more than ten dozen times. Note, however, that these books are frequently informational, such as the law books Mr. Breen consults when he receives a slanderous postcard and the reference books in the Blooms' library (17.1361, 1391). When Bloom imagines the library in his dream home, he mentions only informational books: "the Encyclopaedia Britannica and New Century Dictionary" (17.1523–24). When Joyce uses distinctive typography to remind the reader that *Ulysses* is a print book as well as a novel, he strengthens the association between novels and books as information storage media.

This metamedial strategy is further developed through *Ulysses*'s descriptions of the economic and aesthetic value of books. The books in the Blooms' library are catalogued like valuable commodities: the narration details their color, binding type, and physical condition. The descriptions mimic the language used by booksellers: "tooled binding," "cover wanting, marginal annotations," "S plates, antique letterpress long primer, author's footnotes nonpareil," and so on (17.1365–97). This language is parodic; like most of the language of "Ithaca," it is hyperbolically technical jargon.[90] This language may testify to Bloom's keen eye for valuable books.[91] Given Joyce's own interest in book design, such moments also call attention to *Ulysses*'s dual status as novel and book. As Lawrence Rainey has documented, Joyce "approve[d] the paper, typeface, and page layout for each issue" of *Ulysses*'s first edition, and he "[chose] the color for the cover and even . . . authorize[d] the inks that would reproduce the color."[92] The blue-and-white cover recalled the Greek flag, explicitly forging a metamedial connection between *Ulysses*-as-print-book and *Ulysses*-as-Homeric-retelling. In sum, the economic, aesthetic, and informational registers of print books merge in *Ulysses,* prompting readers to consider how the novel they held, bound in a book, might function similarly to the books described within the narrative.

Like the Mundaneum, then, *Ulysses* is a collection of information vast enough to inspire the belief that it is an archive of culture; also like the Mundaneum, it examines how individual items of information are organized within larger systems and embodied in media. Additionally, both Otlet and Joyce employed the monographic principle as a formal strategy. Joyce tended to compose scenes in *Ulysses* by integrating stand-alone lines, notes, quotations, and other textual fragments that he had collected from sources ranging from literary works to textbooks.[93] *Ulysses*'s many lists best represent this strategy of composition through accumulation, but the work as a whole epitomizes this informational aesthetic. Such literary atomization distinguishes modernist novels such as *Ulysses* from their Victorian predecessors. As James Buzard has argued, nineteenth-century novels "very self-consciously . . . assert[ed] a form of textual organization sharply distinguished from that of the catalogue, the list, the encyclopedia, the state-sponsored blue book or statistical table."[94] In contrast, Joyce experimented with the novel's ability to act as a cultural record formally as well as in terms of content. It is not simply that modernist literature is fragmented, although that descriptor fits. Rather, the assemblage of a larger text out of smaller units made a literary technique out of an information management process. "This may, or may not, be literature," wrote Shane Leslie, one of *Ulysses*'s early reviewers. "It is certainly good cataloguing."[95]

Joyce's cataloguing is particularly evident in *Ulysses*'s lists. Written lists originated in recordkeeping; they manage information by sorting and organizing it. To use Caroline Levine's terminology of forms, lists are bounded wholes, groupings whose "unity joins disparate elements into one."[96] Lists are flexible, in that their defining category may be either straightforward or arbitrary. In the second case, a list's items are united by circumstance instead of inherent similarity. Thus, topics as diverse as "music, literature, Ireland, Dublin, Paris, friendship, woman, prostitution, [and] diet" all fit the category of subjects discussed by Bloom and Dedalus as they walk to Bloom's home (17.12–13).

Given the potential for such expansive inclusivity, a list can act like a microcosm of the Mundaneum: an organizational system capable of classifying and incorporating anything imaginable that fits its category. As Fritz Senn notes in writing about *Ulysses,* "some catalogues are naturally complete, exhaustive. It is possible to list all Greek troops and their navy, every single book in a library or the objects of a kitchen dresser."[97] Joyce's flirtation with archival perfection is strongest in "Ithaca," an episode concerned throughout with the authority, completeness, and accuracy of information. The episode meticulously catalogues the objects in the Blooms' home, at times proceeding drawer by drawer, while the question-and-answer format recalls informational forms such as catechism, catalogue, science textbook, and, prophetically, database.[98] The drive for order comes from Bloom—looking over his books, he thinks about "the necessity of order, a place for everything and everything in its place"—as well as from the episode's unnamed interrogator, who demands a strict accounting: "Catalogue these books," "Compile the budget for 16 June 1904" (17.1410, 1361, 1455).

As these lists gesture toward the scale of modern information, and toward comprehensive organizational systems, they also simulate the cognitive burden of information shock. Joyce's lists are rhetorically varied: some are playful and imaginative, others mechanical and formulaic.

Their sheer comprehensiveness may tire the most ardent reader. The waterhymn—a list of "what in water Bloom . . . admire[d]"—spans more than a page (17.184–85). Its entries range from the mundane ("its weight and volume and density") to the poetic ("its secrecy in springs and latent humidity") (17.200, 208–9). It is a seemingly exhaustive catalogue of manifestations of water: as "vapour, mist, cloud, rain, sleet, snow, hail"; as "icecaps, arctic and Antarctic"; as "waves" and "rivers" and "Artesian wells," to give a limited sample (172.16–17, 192, 187, 222, 205). An overloaded reader may quickly grasp the organizing principle and skip the rest. Joycean scholars are not immune from this effect: Evan Horowitz argues that the lists in "Ithaca" "actively frustrate reading," "demand[ing] . . . to be skimmed—if not skipped altogether."⁹⁹

This compulsion to skim or skip is produced by the disjunction between the conventions of novel reading and list reading. Although one or two dozen entries in a list is not a prohibitive quantity, a list situated within a novel frustrates linear novelistic reading and precludes the random access one would expect in an informational text. *Ulysses*'s lists generally lack any organizing principle beyond the category. Even lists whose contents lend themselves to alphabetization tend to be ordered randomly. The list of attendees at a wedding is amusing, if repetitive: more than thirty entries, each a combination of a woman's name with a plant-related surname (e.g., "Mrs. Maud Mahogany, Miss Myra Myrtle, Miss Priscilla Elderflower" [12.1273–74]). One might reorder this list without much effect. Explicitly informational genres require far more internal structure. Imagine the wedding list as a telephone book; it would be useless. When a novel such as *Ulysses* simulates the experience of information shock, this is not only a matter of content—"an excess of knowledge that the text demands and that the reader does not have" or "an excess of information and sensory data that the text presents and the reader has inadequate resources to manage," as Leonard Diepeveen describes in his seminal account of modernist difficulty—but also of a mismatch between the organizational frameworks and consequent reading strategies of novels and those of informational texts.[100] Information professionals deemed the book poorly scaled to manage information flows, but Joyce showed how the experience of scale could be re-created within the book's limited space.

AGAINST TOTALIZATION: THE LIMITS AND POLITICS OF "UNIVERSAL" SYSTEMS

Despite Joyce's enthusiastic cataloguing, *Ulysses* undermines the possibility of a perfectly complete, accurate, or organized collection of information. One way in which this occurs is metamedial, as *Ulysses* acknowledges its material limits. In "Ithaca," the episode so preoccupied with infinity, Bloom considers the finitude of paper. Bloom's attempts to calculate the size of the universe are frustrated by the amount of paper necessary for the task:

> Why did he not elaborate these calculations to a more precise result?
>   Because some years previously . . . he had learned of the existence of a number computed to a relative degree of accuracy to be of such magnitude and of so many places . . . that, the result having been obtained, 33 closely printed volumes of 1000 pages each of innumerable quires and reams of India paper would have to be requisitioned in order to contain the complete tale of its printed integers. (17.1070–77)

"Innumerable quires": a prohibitive, perhaps impossible, quantity. The first edition of *Ulysses* contained 732 pages and was "as large as a telephone directory or a family bible"; as Joyce composed the novel, he described his materials as a "bewildering mass of papers."[101] The description of "innumerable quires and reams" metamedially references *Ulysses*'s considerable size, but it also emphasizes that even this unusually bulky, information-dense novel is limited by the storage capacity of its medium. The reference to India paper is also significant. Sturdy and very thin, India paper was a practical choice for a reference text (or the Oxford University Press's popular India-paper Bibles) but unnecessary for most literary material. The reference to the thirty-three volumes of India paper differentiates *Ulysses*—whose first edition was printed on handmade Holland paper, vergé d'arches, and linen paper—from actual reference works such as the 1911 edition of the *Encyclopaedia Britannica,* whose twenty-nine volumes were printed on India paper.[102]

Joyce's deeper claim is that even the vastest collection of information cannot match the scale of what lies beyond quantification. "Ithaca" presents the ultimate subject for totalistic accounting: the universe. As Bloom points out constellations, the narration stresses the limits of measurement. The Milky Way is "infinite" (17.1043). Stars are "wanderers from *immeasurably* remote eons to *infinitely* remote futures" (17.1053–54, emphasis added). The narrative lens then shifts from telescopic to microscopic, culminating in a description of blood cells dividing into increasingly smaller parts "till, if the progress were carried far enough nought nowhere was never reached" (17.1067–69). The smallest objects expand into a state of unparsable negation. Here one might be reminded of *Tender Buttons,* whose determined accounting so often lapses into similar negation: "nothing elegant," "nothing flat," "nothing, nothing at all," and so on.[103] Brian Cosgrove argues that the "problem explored in 'Ithaca'" is that "any schema we devise (linguistic, scientific, philosophical) cannot hope to accommodate in its totality a reality which in its 'infinite plentitude' . . . eludes our schematic categories, our conceptual reductions."[104] Paper is too limited for Bloom's calculations, but the microscale of microform could not record the macroscale of a universe where a "single pinhead" contains "incalculable trillions of billions of millions of imperceptible molecules" (17.1063, 1061–62). Otlet and La Fontaine dreamed of an archive that could organize the world; *Ulysses* takes on the universe, only to record the impossibility of perfect accounting.

Ulysses also emphasizes the incompleteness of collections. Earlier I quoted Senn: "It is possible to list . . . every single book in a library or the objects of a kitchen dresser." Presumably the list of books in the Blooms' library is complete (17.1361–98), but the description of the contents of the Blooms' kitchen dresser is not. The long passage details more than fifty items on the dresser's three shelves, from "five vertical breakfast plates" and "a chipped eggcup containing pepper" to "an empty pot of Plumtree's potted meat" and "a small dish containing a slice of fresh ribsteak" (17.298, 302–3, 304, 316–17). Despite the volume of information, omissions remain in Joyce's description. As if the narrator is exhausted after describing the first two shelves, the third is simply described as holding jars "of various sizes and proveniences" (17.318). Joyce could have provided a precise description of every object in the dresser but failed, or refused, to do so. Such omissions chime with *Ulysses*'s acknowledgment of its own necessary incompletion: set entirely on June 16, 1904, it reminds the reader that "every life is many days, day after day" (9.1044).

Most fundamentally, Joyce problematizes the philosophical grounds on which modern information systems were founded by showing order to be always on the brink of collapse. Despite Bloom's musings on "the necessity of order," he is beset by disorder: his books are upside down and unalphabetized, and he is injured by the disarray of his furniture (17.1410, 1269–90, 1358–60). The novel's errors and glitches are another symptom of disorder: *Ulysses* is concerned throughout with the complications that attend information transmission and storage, such as telegraphic typos, static in gramophone recordings, and nonsensical text in newspapers.[105] On the level of the novel as a whole, *Ulysses* destabilizes its mythic correspondences. The Gilbert and Linati schemas are only partially consistent. This is similarly the case with the Homeric parallels, which offer not so much correspondences as ironic points of comparison. Rather than a new mythos, *Ulysses* offers the mock-heroic: Odysseus reduced to a slightly ridiculous Dubliner, subaltern rather than hero; faithful Penelope an adulterer. These modes of ordering do not provide a new metanarrative. Instead, they strain at the boundaries of ordering systems. The network of Greek allusions is undermined on the first page, as Buck Mulligan teases Stephen Dedalus about the absurdity of his surname (1.34).

This collapse of organizational systems has implications for postcolonial conceptualizations of identity and, more broadly, for understanding how information systems benefit from and perpetuate power. As Enda Duffy and Vincent Cheng have argued, Joyce's works habitually take issue with essentialist constructions of Irishness that replicate the strict Irish-English dichotomy that was central to colonialist discourse.[106] *Ulysses*'s three protagonists resist simple classification along the lines of national categories.[107] In "Cyclops," an episode in which list items frequently challenge list categories, the boundary between Irish and non-Irish is destabilized by a catalogue of "Irish heroes and heroines of antiquity" that includes not only actual Irish heroes ("Cuchulin, Conn of hundred battles, Niall of nine hostages") but also non-Irish figures ("Patrick W. Shakespeare, Brian Confucius, Murtagh Gutenberg") (12.176–77, 190–91). The list is humorous but also performs political work, challenging the xenophobia of the Citizen, the Irish nationalist whose attack on Bloom stems from his belief that there are only two categories: Irish and Other.[108] The Citizen himself is the most easily categorized of *Ulysses*'s characters, existing as a list of Stage Irish attributes: "redhaired," "freelyfreckled," wearing "a loose kilt," etc. (12.153, 169). These examples caution against the dangers of a worldview based on rigid, reductive classifications.

The target of Joyce's critique in these cases is Irish nationalism rather than British imperialism, but these examples speak to the tendency of systems that organize information to reify privilege and to the restrictive constructions of identity on which hegemonic power relies. In the late twentieth and early twenty-first centuries, professionals in library and information science have worked to mitigate the inequity and marginalization that result from the imposition of supposedly universal classification systems. In her seminal study, Hope

A. Olson argues that "standards homogenize the results of cataloguing and, thus, impose a universal language on diverse contexts," with the result that many users are misrepresented, impeded in their access to information, or otherwise disempowered.[109] Classification systems such as the Library of Congress Subject Headings can perpetuate bias, as with the now-obsolete heading "Yellow Peril" for the topic of Asian immigration.[110] They reinforce the epistemological foundations of racism and colonialism by furthering the notion that, as Edward Said put it, "a specific body of information" about subaltern groups of people "seem[s] to be morally neutral and objectively valid."[111] Safiya Umoja Noble writes that Eurocentric perspectives "dominate the canons of knowledge" in the United States and Europe because "knowledge management reflects the same social biases that exist in society."[112] Activists have created alternative information frameworks, such as the Brian Deer Classification System and Ngā Upoko Tukutuku (Māori Subject Headings), to counter the perspectives of Eurocentric systems by using indigenous organizational concepts.[113]

If one attempts to categorize *Ulysses* within Otlet's Universal Decimal Classification, Olson's argument that "universal solutions are not viable options" becomes apparent.[114] "Literature" is one of the UDC's ten base categories. The first subdivision for literary subjects proceeds along national lines: literature's eight subcategories in early editions of the UDC were English, German, French, Italian, Spanish, Latin, Greek, and "other literature." A literary text could be further classified by eight choices of genre: poetry, drama, fiction, essays, speeches, letters, satire, and miscellaneous.[115] Even leaving aside the contentious question of *Ulysses*'s genre, what national subcategory would it best inhabit? Joyce's Irish origin should place *Ulysses* in the miscellaneous "other literature" category, a classification that downplays Ireland's rich literary history (and those of all other nations beyond the seven listed European countries). This heading also fails to register *Ulysses*'s international contexts: it was written in Trieste, Zurich, and Paris; published in Paris; based on the *Odyssey;* and so on.

My aim is not to criticize Otlet and La Fontaine. In fact, the pair acknowledged that the best system would always be an approximation. La Fontaine stated that universal classification systems did not map well onto local and national ontologies.[116] Moreover, lest we be tempted to overpraise Joyce's critique of totalizing systems, we should recall that he invoked universality throughout *Ulysses* and especially in *Finnegans Wake*. He also developed his own personal organizational systems, such as the twenty subject headings in what scholars call his Subject Notebook, where he assembled notes on various topics for *Ulysses*. These headings include character names, topics such as "Art" and "Theosophy," and, pertinent to this discussion, the category "Jews." The existence of this last category carries the danger of essentialism, reproducing, albeit to a lesser degree, the Citizen's view of Jewish people as outsiders, as classifiable others. Wim Van Mierlo, however, notes that the other subject heading relating to the "political canvas of *Ulysses*" is "Irish," and he argues that the notebook juxtaposes "Jews" and "Irish" meaningfully: "Joyce for the first time in the notebook used a verso page . . . for the notes under 'Irish' and placed them alongside 'Jews' on the accompanying recto . . . , suggesting, as elsewhere in his work, a link between the plights of the two people."[117] Even as Joyce developed his own subject heading system to write *Ulysses,* the text acknowledges how totalizing systems impose power disparities and explores their limits and the circumstances under which they break down.

MEANINGFUL SYSTEMS: INFORMATION, THE NOVEL, AND THE BOOK

Despite *Ulysses*'s negation of totalizing, universal systems, the novel does not abandon the possibility of strategic information management. Instead, it carves out a middle path between ceding all comprehension of information to a rigid, highly mediated system and succumbing to information shock. This middle path is grounded in Joyce's attentiveness to the ways in which the print book organizes the text it contains.

Van Mierlo's argument that Joyce suggested parallels between Jewish people and Irish people by placing these categories across either side of a page spread underscores the importance of how information systems are situated in media. It also demonstrates the significance of the order and organization that books impose on their contents.

Ulysses does not anarchically celebrate chaos. Instead, readers must move among different ordering systems—a reading process that mimics the operation of one of *Ulysses*'s keywords, *parallax*. Parallax is the difference or displacement that appears when an object is seen from two different points of view; the synthesis of these two positions creates depth perception. Parallax has provided a popular critical metaphor for *Ulysses*'s narrative structure and consequent reading strategies.[118] Rather than debate the finer points of critical readings of parallax in Joyce, I want to make the point that when Joyce's perspectival play occurs at the level of system, it suggests that the best approach to large-scale information is to move, flexibly and continuously, among multiple organizational schemas. No single frame or schema accounts for every detail and datum in *Ulysses;* taken together, however, the various frameworks allow readers to make sense of the work. By extension, Joyce suggests that what is needed to properly understand information in the world—not only to manage it but also to theorize it in its complexity—is multiple organizational systems.

Ulysses's lists, for example, may compel readers to skim them, but they also invite close, contemplative reading. The waterhymn becomes poetic elaboration, a neo-Homeric catalogue. The passage is compelling for the same reason it is challenging: because of its unpredictability. The list might include any aspect of water in any order. In terms of the information theory formulated by Claude Shannon and Warren Weaver, this unpredictability means that there is maximum information in each item.[119] The waterhymn's shifting rhythms evoke water's slow swells and quick breaking: consider the appropriately slower pace of "its imperturbability in lagoons and highland tarns" compared to the rapid progression of "its violence in seaquakes, waterspouts, Artesian wells, eruptions, torrents, eddies" (17.200–201, 205–6). As a whole, the passage is daunting, but the lyricism of individual lines rewards close reading. Grasping the list's gist is a quick task, diametrically opposed to a slow, deliberate working through of its components. Both are necessary because the waterhymn's operation is meaningful in the interplay between these modes. This is not merely "good cataloguing." It is exhaustive, excessive cataloguing as poetics. It is also a model of reading as oscillation among different perspectives. As we best make sense of Joyce's lists by taking into account both gestalt and individual elements, we best read *Ulysses* by alternating among schemas, myths, literary allusions, and other systems. Bloom is a psychologically believable character; he is Odysseus; he is an anti-Odysseus; he emblematizes digestion; and so on. He is all of these simultaneously. There is no all-encompassing ordering system in *Ulysses;* the multiplicity of frameworks offers opportunities for contextualization without asserting totality.[120] The text works against the mediating logic and organizational stability of modern information management, prompting readers to grapple with the forms and signifying systems that make information meaningful. As *Ulysses* emphasizes information's forms by foregrounding its multiple frameworks, as well as its print artifactuality, it shows information to be emergent and becoming.

This is an appropriately novelistic philosophy of information. As I discussed in the introduction, theorists frequently define the novel not as a set of formal characteristics but as an investigation into the nature of genre categorization. Michael McKeon writes that "the novel crystalizes genreness, self-consciously incorporating, as part of its form, the problem of its own categorical status. What makes the novel a different sort of genre may therefore be not in its 'nature' but in its tendency to reflect on its nature."[121] Questioning how genres are classified, the novel foregrounds how patterns are recognized and meaning is produced. Additionally, novels are always multiply framed, setting a tension between organizational systems at the heart of the genre. At a minimum, a novel is organized by the linear time in which the plot occurs and the order in which the narration recounts these events. Modernist novelists from Ford Madox Ford to William Faulkner emphasized the disjunction between these frameworks. They also innovated with multiple perspectives, indicating how the same object may appear from different points of view. In *Mrs. Dalloway's* skywriting scene, the Londoners viewing the airplane disagree not only on the letters they are reading but on the significance of the skywriting (an advertisement, if a wonder, to most onlookers, but "exquisite beauty" to Septimus).[122] There is no final consensus on the skywriting's meaning. Its significance arises from the cumulative amalgamation of perspectives, and from their divergence from one another, with each no less meaningful for being partial or irreconcilable with the others.

In *Ulysses,* these insights are reinforced through Joyce's metamedial emphasis on the book. Print books also form data into information through their organizational structures. A book contains multiple ordering systems: the arrangements and divisions of text (into sentences, chapters, and sections); the space of each page, structuring the content in two dimensions; the ordering of pages into the codex, structuring the content in three dimensions; etc. When a reader holds a book, its dimensions communicate information about the amount and scope of information it contains. Likewise, the order in which the book is divided and bound communicates information about how each element relates to the others. These physical cues aid in understanding content. Readers tend to remember where information is located spatially in a text, they construct cognitive maps of texts based on the contextual cues of textual media, and they better remember the chronology of narrative events when they are reading in books compared to electronic contexts.[123] Volumetrically, print books provide "haptic and tactile feedback," giving readers a "tactile sense of progress."[124] Books impose multiple organizational structures, making these structures apprehensible to readers with a legibility and immediacy that most modern information management systems were designed to avoid. The hypermediacy of the book in a metamedial text such as

Ulysses is the conceptual opposite of the transparent mediation of the systems used by institutions such as the Mundaneum.

This hypermediacy renders visible the novel's conventional uses of the print book. As such, it illustrates another insight: that novels, like media and formats, are theorized through their disruptions of conventions and default forms. As I indicated earlier, I am aware that this chapter rests on a precarious assertion: that we can theorize the modernist novel writ large through an examination of *Ulysses*. Here, I defend that assertion with another claim: it is precisely *Ulysses*'s singularity that makes it a useful case study for theorizing the novel in the context of media culture. Exemplarity is a problem that literary studies as a whole has yet to resolve. Concerns about canon formation and the selection of representative texts are one reason why digital humanities methods have proven so compelling, promising to identify patterns across tens of thousands of texts.

For the theorization of the novel, the selection of exemplars has additional complications. The novel's history is built on two impulses in tension with one other: to model the genre on existing forms as a means of establishing legitimacy and to search for new forms in a quest for literary novelty. As J. Paul Hunter describes the novel's early history: "because of its obscure beginnings and insecure place in the hierarchy of literary forms and kinds, the novel historically has asserted itself by comparison to more established forms." At the same time, these works evinced a "desire to make clear that the novel was new and different, needing to . . . show its distinctiveness from other forms of fictional narrative."[125] Both impulses establish the novel as a form that perpetually problematizes its formal coherence. While some novels seem more typical than others, all case studies must be singular ones, given "the way the novel has been theorized, for most of its history, as a genre singularly deficient in generic identity."[126] To theorize the novel is to theorize what *is* novel, and that novelty is created partly by self-conscious attempts to work against extant conventions. *Ulysses*'s extreme stylistic innovation and concerted encyclopedism may make its classification as a novel debatable, but I am suggesting that this is a problem not so much unique to *Ulysses* as endemic to the study of the novel. As Hunter writes, "not everything in novels . . . goes the way critics and critical theorists think it should, and some of the 'failures'"—and, we might add, the limit cases—"are characteristic of the species, even definitive."[127]

Metamedial novels such as *Ulysses* prove the rule of both novel and printed page in the original sense of that phrase: they test and violate the rules that govern conventional typographical page layouts as they strain against conventional articulations of novelistic form. They also make those rules more visible precisely by demonstrating that a convention is only one of many possibilities. Similarly, the model of information Joyce creates in *Ulysses* may not epitomize the dominant novelistic response to information shock, but it does model one approach. This is the work of metamedial novels during the modernist period, as again in the contemporary moment: to highlight the print book's conventions and possible uses in order to explore how the book's and the novel's affordances as information media interoperate. The novel's many frames—formal, conceptual, narrative, generic, and medial—offer readers many systems for understanding content. Meaning lies in excess of any single system, but all systems are grounded in their medial embodiment. This is the condition of the novel; *Ulysses* makes clear that this is also the condition of information in the world.

The phrase "too much to read" recurred like a refrain in early-twentieth-century texts, invoked by readers, novelists, and information scientists alike. Otlet and the period's other information workers admirably met the challenges of a vast and variegated media culture, and of unprecedented volumes of information, with optimism and pragmatism. They developed comprehensive and practical ways to cope with information abundance. Yet they were not the era's only information architects. Novelists also considered the issues raised by the scale of information, and they offered their own solutions, grounded in their engagement with the book. To read the modernist novel through the lens of modernity's information history—and vice versa—is to complicate critical accounts of both. The standard narrative would have it that the modernism of Joyce celebrated fragmentation and chaos while the modernity of Otlet celebrated order and systematization. This chapter has demonstrated instead that both speak to experiences of textual proliferation and information shock, and both were fascinated with the tension between order and disorder, monographic fragment and system.

Since its inception, the novel has reflected on epistemology: on the nature of what can be known, on the relationship between fiction and reality, and on how the artifice of narrative might capture or distort the world. With the advent of modern information culture, these broad generic concerns narrowed to a focus on the structures, limits, and representation of information. Joyce and his contemporaries explored the cognitive burdens of information abundance as well as the methods, and consequences, of its mediation and management. Novelists' emphasis on the ways that narrative, medial, and organizational systems of order interact clarified the stakes of information management's movement from books to modular information systems. The mediating interfaces so integral to these systems worked on a practical level to manage scale; on a philosophical level, however, they concealed the actual forms that information took. By foregrounding the book's organizational hypermediacy, *Ulysses* contributed to the conversation about information scale, insisting that the apparent perfection of a system such as the Mundaneum's is always itself a fiction, a practical abstraction rather than an accurate model of information in its full complexity.

"You must pick and choose [what to read] with deliberate purpose, or you must flounder"; modernist novels instead urged readers to develop individual interpretative frameworks rather than cede the agency of understanding to a reader's guide, card catalogue, or other

information interface. By modeling information abundance, modernist novels served as a training ground where readers could practice
making sense of challenging quantities of information, reflecting as
they did so on the ways in which meaning is made and the role media
play in shaping interpretations of data. These novels illuminate the
importance of the book for early-twentieth-century conceptualizations
of information and mediation: whether as a medium that foregrounds
its organizational structures or as an emblem of the older information
management techniques against which modern information scientists
defined their new systems, the book remained central to the discourse
of information.

*Information
Shock*

Form in the Cloud

Computational Mediation and the

Contemporary Novel

All that information ... looms over us, not quite visible, not quite tangible,
but awfully real; amorphous, spectral; hovering nearby, yet not situated in
any one place.

—*James Gleick,* The Information

To speak of writing ... is to speak of a spasm or explosion. ...You do not read
writing; you cannot take in the mass of texts in the world. ... The writings
exceed you, they overwhelm you, and they bury you. ... Our entire species is
devoted to producing greater and greater explosive spasms of overwhelming
printed matter. Is this not the network? Is this not the web? Not texts, not
writing to be read, but writing as massed marked detritus.

—*Sandy Baldwin,* The Internet Unconscious

TL;DR

Matthew Jockers begins his digital humanities treatise *Macroanalysis*
with reference to the transformative potential of twenty-first-century
information scale. He quotes a *Wired* magazine headline from 2008—
"Data Deluge Makes the Scientific Method Obsolete"—and proposes
that literary scholarship must undergo a similarly radical shift. Big
Data has revolutionized the sciences, he claims, and it is time to rev-
olutionize the humanities: "we have reached a tipping point, an event
horizon where enough text and literature have been encoded to both

allow, and, indeed, force us to ask an entirely new set of questions about literature and the literary record."[1] Jockers's assertion that the pace of human reading does not scale to the archive of literary history is a familiar claim in the digital humanities. According to this argument, without the assistance of technology, no individual scholar can engage in what John Unsworth calls "reading at library-scale," the kind of reading necessary to model, as Franco Moretti puts it, literature's "large mass of facts" into "a collective system."[2] Although many practitioners of the digital humanities (Jockers included) stress the need to contextualize macroanalysis with careful attention to the texts being studied, the discipline is often framed as discarding close reading in favor of distant reading.

The volume of information accessible in the twenty-first century, combined with the technological processes that manage it, has altered dominant paradigms of reading and information navigation. While reports on how much people are reading have been optimistic in recent years, the concern that people are reading less carefully, and less often for pleasure, continues to drive analyses of reading habits. Online, targeted search and aleatory browsing are behaviors better adapted to the magnitude of available information. Even academics "power-browse" online, typically reading only the first few pages of journal articles in electronic reading environments.[3] In institutions such as businesses and governments, information processing has superseded reading: Big Data analytics and the algorithmic processing of information, in quantities far surpassing even the idea of library scale, have supplanted human decision making. The early-twentieth-century complaint of "too much to read" has morphed into the Internet-era quip "too long, didn't read"—abbreviated to "TL;DR," as if the full (brief) phrase is itself excessive.

Debates about digital media's cognitive and social impacts frequently construct literary reading as the polar opposite of the computational navigation of large quantities of textual information. According to the co-founder and CEO of Spritz, a speed-reading app for mobile media, "you wouldn't really want to read classic lit[erature] or Shakespeare" on Spritz.[4] Literary texts, he assumes, exist to be read slowly and deliberately. This expectation is heightened for novels, which have become closely associated with print-centric, literary models of reading. "The literary novel," opines Sven Birkerts, "represents reading in its purest form."[5] As I wrote in the introduction, while I do not wish to reify the privileging of the novel as a site of deep reading, the claim that novels by writers such as Charles Dickens, Henry James, and Jane Austen cultivate an attentional mode at odds with the attentional modes cultivated by digital media has become ubiquitous in accounts of digital information culture's effects on attention and cognition.[6]

This chapter considers how the study of the print novel has the potential to correct problematic assumptions about digital information scale and its management. (Historians of the novel will have already spotted the irony of Birkerts's claim: for many years, novels were viewed as the epitome of careless or extensive reading—as a direct threat to deep attention.) Given the novel's centrality in critiques of digital mediation and long history of contemplating the nature and relevance of its own information scale, it is the ideal literary genre to study in the context of contemporary information culture, and it offers a strategic vantage point on the information management practices of the Internet era. I examine how and why novelists have grappled with the implications and aesthetics of contemporary information scale by foregrounding the formal structures of the print book. In the so-called post-print era, the book is a legacy medium. Its storage capacity is dwarfed by that of digital media, and it differs fundamentally in the ways in which its users access and navigate information. Yet the print book can highlight how the framing and contextualization of data generates meaningful information. As such, this chapter argues, it offers a significant alternative to the decontextualization that occurs with computational mediation.

Computational mediation——the use of interfaces and automated, algorithmic systems to retrieve, display, and analyze data—is the primary means of managing the contemporary information landscape's staggering scale. As of 2014, there were more than a billion websites online.[7] Search engines field billions of user queries daily.[8] Estimates of global data production vary, but reports place the total at around 2.5 exabytes (2.5 billion gigabytes) per day. (By comparison, in 2007, experts estimated that about 5 exabytes of data were produced each *year*.)[9] The magnitude of this quantity is difficult to comprehend. One blog post by IBM concretized it by explaining that 2.5 exabytes have "18 zeroes."[10] Historian of information James Cortada's explanation makes reference to the space occupied by books: "one gigabyte has been likened to ten yards of books on a shelf. Now multiply that by 2.5 billion."[11] Put another way, it is the equivalent of 250,000 Libraries of Congress—daily.[12] The majority of these data exist in digital formats, and the majority are unstructured, lacking the formatting that would make them easily organized or sorted. In its scale, as in its mediated and unstructured nature, digital information frustrates attempts to understand its form.

The novels I focus on in this chapter—Mark Z. Danielewski's *Only Revolutions* (2006) and Matthew McIntosh's *theMystery.doc* (2017)—explore the role of form in the digital information ecology by bringing metamedial attention to the print book's form. The dominant metaphors of digital information—the Internet as a cloud, hardware as a black box—indicate its opacity. In contrast, the print book makes its formal organization immediately apprehensible. The relationship of each part to the whole is unchanging: a line on page 1 remains the same set distance from a line on page 10, whether measured in terms of words or by the distance between bound pages. Formally, the book conditions a wholly different set of expectations for the information that users should have about information. *theMystery.doc* and *Only Revolutions* examine the tension between the scale, density, and complexity of novelistic information on the one hand and the legible immediacy of the print book's organization on the other.

Developing strategies for the representation and critique of information scale and its mediation, *theMystery.doc* and *Only Revolutions* raise crucial questions for critical information studies and book studies. How do interfaces—particularly the search interfaces used to navigate information online—shape conceptualizations of information, and what are the consequences of this mediation? What role can books play in managing or conceptualizing unprecedented quantities of information? Because computational interfaces provide "neat and compact presentation[s] of a messy and sprawling world," the media and mediations that give form to data greatly impact the way in which those data are understood to be meaningful.[13] Wedding an emphasis on the book's material form with the novel's longstanding interests in information culture and scale, metamedial novels such as *Only Revolutions* and *theMystery.doc* create a rich framework for answering these questions. Although these novels' dominant techniques are diametrically opposed—*Only Revolutions* is governed by tight constraints and a program of literary compression, while *theMystery.doc*'s sprawling 1,600-plus pages constitute a maximalist assemblage—both insist that what matters most about large quantities of information is how that magnitude is made meaningful as it is limited and framed. As they strive to represent the vast scale of information culture within the limited spaces of novel and book, they prompt readers to reflect on how they mediate information in a TL;DR world.

FORM, REPRESENTATION, AND THE PRINT BOOK IN THE AGE OF GOOGLE SEARCH

Google has had arguably the largest impact of any organization on contemporary information culture. It has shaped how questions are asked and answered online, how the world is mapped and navigated, and how digital searches are monetized. From media studies to critical algorithm studies to feminist information studies, much scholarly analysis of digital culture has focused on what Siva Vaidhyanathan terms "Googlization" and its influence on the politics of search.[14] Because Google Search's algorithmic operations and interface aesthetics typify wider practices in information management, their study illuminates the ideologies underlying computational mediation—ideologies crucial for understanding not only the more literary aspects of information culture (that is, its impact on reading and authorship) but also the broader role that form plays in communication. The Google brand has become shorthand for search, "synonymous for many everyday users with 'The Internet' itself."[15] From Google's development of a business model "designed to scale to no boundary" to its corporate mission of "organiz[ing] the world's information," Google has had a profound impact on the cultural imaginary of twenty-first-century information scale.[16]

Computational mediation is inextricable from digital information culture. It occurs when Google returns results from a search query or when Facebook generates a personalized newsfeed. Mediation is ubiquitous in computing culture, bolstered by the long-standing design principle that interfaces should be user-friendly.[17] Search engines such as Google's are a pragmatic concession to the scale and complexity of information available online. The Internet is an enormous assemblage of information, infrastructure, and users. In the early days of the World Wide Web, the Internet was likened to a library grown out of control—the modern-day equivalent of Borges's infinite Library of Babel, for instance, or, in the much-quoted words of the librarian Michael Gorman, "a huge vandalized library" where "someone has destroyed the catalog and removed the front matter, indexes, etc. from hundreds of thousands of books and torn and scattered what remains."[18]

In these print-centric descriptions, as in descriptions of online information as a flood or a tsunami, the Internet functions as both the fulfillment and the negation of Paul Otlet's dream (discussed in chapter 1) of a total archive storing all of the world's information.[19] The fulfillment lies in the Internet's breadth and scope, which users often (wrongly) equate to a compendium of all extant information. It is also the negation because Otlet's totalizing model of information management relied on a regimented, hierarchized classification system. The Mundaneum's dominant visual metaphors, the sphere and the subdivided circle, symbolized a complete and coherently structured archive. The standard iconographic depiction of the Internet as a cloud suggests the opposite: a formless mass. The cloud represents the fact that users do not need to be aware of the operating details of the resources they use online. In fact, the Internet is highly structured; if we perceive it to be formless, this is due both to deliberate interface design and to the overabundance of formal components that constitute the global Internet.

The Internet has multiple levels of organization: semantic order (the relationship of information on one webpage to that on other pages on the same website or elsewhere on the Web), the internal structures of the databases that store this information, the global dispersion of Internet infrastructure (such as underwater cables, server farms, and network access points), the structures of screen displays and HTML source code, and so on. The Internet is also a mélange of organizational configurations: at the internet layer, router configuration may be rhizomatic, whereas, at the application layer, domain names are hierarchical. To give an example of the daunting task of understanding the Internet's organization: how might one visualize Internet Protocol (IP) addresses? IP addresses are assigned to each device connected to a network; by one estimate, these devices number around 460 billion.[20] Should they be represented via their geographic location or by device type? Should they be mapped according to number patterns or their allocation to nations and corporations? Should the visualization center on IP version 4, IP version 6, or both? Little wonder that so many visualizations of the Internet take the form of networks of points so small and numerous that they cluster into masses, indistinct blurs that represent the limit of representation. This is what Alexander Galloway has termed "the dilemma of *unrepresentability* lurking within information aesthetics": "An increase in aesthetic information produces a decline in information aesthetics."[21] Pushed to its limit, a network becomes a cloud.

Google Search attempts to represent neither the totality of information online nor the totality of information on any one topic. Instead, it retrieves a list of results, decontextualized from systems of meaning. Google Search epitomizes the guiding assumption of twenty-first-century information culture: that information cannot be made sense of as a whole. According to this assumption, the quantity is too great to be represented, much less conceptualized; it defies aestheticization and understanding. Enter the algorithm. When I type a query, Google Search uses natural language models to parse my intent and then searches its web index for relevant matches based on keywords and their synonyms. This index is constantly updated as Google's web crawlers track hundreds of billions of pages.[22] These search results are ranked via their PageRank score, with higher-scoring webpages receiving more prominent placement in the list. Google Search considers more than two hundred elements in the ranking process, such as the frequency with which keywords appear on a webpage and how long the site has existed online. It also assesses the more subjective measure of website quality, which Google determines by "counting the number and quality of links to a page to determine a rough estimate of how important the website is."[23] As Craig Dworkin has described, the movement toward nonhuman, automated web search, epitomized by Google Search, represents a reorientation from the early years of the web, when users discovered new content via thoughtfully curated lists (from top-ten lists of favorite movies to academic catalogues such as Alan Liu's *Voice of the Shuttle*).[24] In stark contrast, the algorithms that have dominated online search since 2007 "obscur[e]" "the substantial human labor involved in their creation and deployment" as "algorithmic invisibility easily slides into a presumed neutrality."[25]

The algorithmic mediation of information in online search, and its presumption of neutrality, has serious consequences. As Ed Finn has argued, Google is "a company, indeed an entire worldview, built on an algorithm."[26] Key to this worldview is the belief that algorithms substitute computational accuracy for human bias. Engineers make adjustments to search mechanisms, but the company policy is to let the algorithms determine search rankings and results without human manipulation except "in very limited cases."[27] To override the algorithmically generated results "would contradict the culture of the company, which [is] committed to organizing and presenting information based on math, rules, and facts, not on opinion, values, or judgment."[28] As Vaidhyanathan writes, however, Google's "biases . . . are built into its algorithms," "affect[ing] how we value things, perceive things, and navigate the worlds of culture and ideas. . . . We are folding the interface and structures of Google into our very perceptions."[29]

The social consequences of algorithmic culture's presumption of neutrality have been profound.[30] Algorithmic analyses of Big Data are ubiquitous in corporate and legal decision making, where they often incorporate biases. These biases, like the operations of the algorithm, are largely invisible to those affected. Algorithms have played an integral role in many problematic cases, from credit scores damaged when false records of criminal activity were erroneously added to files; to longer sentences imposed on defendants who are African American or poor or both, based on implicitly biased statistics; to skewed perceptions of social intimacy, based on the positioning of posts in Facebook feeds.[31] As Safiya Umoja Noble has documented, online search, too, has reinforced harmful stereotypes about race, gender, and class.[32] Big Data has been sold as a revolution. The sheer scale of information being produced and captured, combined with the introduction of sophisticated algorithmic analyses, makes it possible to answer new questions. Every revolution has its casualties, however, and one consequence has been the veiling of bias under the guise of technological efficiency.

To examine the multiple levels of mediation operant in digital culture is to recognize that representation—and its negation and obfuscation—is inextricable from twenty-first-century information scale. The historically unprecedented quantities of information available online test the limit of aesthetic form to represent magnitude and complexity. The forms one does encounter online, such as those of search interfaces, take the place of the sublimely incomprehensible underlying information. They conceal the operation of the algorithms (and human-encoded biases) that power them, and they condition the user's ideas about the Internet as well as information more broadly. These are the ideologies of today's information culture: that information retrieval should appear to be quick and simple; that users do not need to contextualize individual items of information within broader systems of meaning; and that, consequently, user agency in locating and conceptualizing information is unimportant.

Consider the Google Search interface. Its simple design distinguished Google from its competitors when the service was launched, and it remains integral: the logo, a query box, and a generous expanse of uncluttered white space. Salient to its early users were not only the "superior search results" but also the fact that "Google spared them the irritating pop-ups, flashing banners, and other mutating forms of advertisements that at that time were competing in an escalating arms race for a visitor's attention on the web."[33] When I visit Google.com more than two decades later, there is little to draw my eye away from the large multicolored logo in the center of the screen and the search box that sits below it. The screen is primarily monochromatic and displays only about a dozen words. With the aim of creating a "strikingly simple" user experience, Google's design principles belie the complexity of the underlying search mechanisms.[34]

This disjunction between surface and system is furthered by the search interface's incorporation of print culture aesthetics. Google's first public logo used the Baskerville Bold typeface. John Baskerville designed the typeface in the 1750s; by his death in 1775 it had become widely imitated. Baskerville was revived in the 1920s and has persisted in digital settings. According to the typography historian Alexander

Lawson, "by every measure, Baskerville's types have demonstrated universal appeal. The proof of this is their present availability throughout the world in the form of single types for hand-composition and for all of the typesetting machines from hot-metal to cathode-ray-tube."[35] Google's decision to use a typeface dating back to the eighteenth century linked the corporation with the long history of print. Given that, as we have seen, the Internet has been praised and critiqued in comparison to print culture, the use of Baskerville Bold granted Google some of print's cachet while also gesturing to the way in which print's conventions have been routinely updated to suit new technologies. At the time of this writing, the Google logo uses the custom typeface Product Sans; its designers cite "schoolbook letter-printing style" as an inspiration.[36] Although Google has created information ideologies wholly separate from those of the print book, print culture has lingered in its visual aesthetics. Google Search's design suggests that to inhabit this virtual space is to enjoy a congenial, relaxed atmosphere—the affective antithesis of the information overload and corporatization typically ascribed to digital culture. When I type a query into the search box, for instance, I can select "Google Search" or the "I'm Feeling Lucky" button; the latter, when moused over, suggests that I might also be feeling "wonderful," "curious," or "playful." Google describes its design in similarly positive, whimsical terms: "quirky," "friendly," "approachable," "unconventional."[37] These words humanize the search experience.

Noticeable in the presentation of search results, however, is how little energy or effort the human user needs to expend in seeking results or reflecting on the retrieved information. If I type "future of the novel," Google returns more than 500 million hits. The ten results that PageRank deems to be most relevant range from a *Guardian* piece called "The Novel is Dead (This Time It's for Real)" to an Amazon listing for Anaïs Nin's *The Novel of the Future* (1968). Users typically do not click past the first ten results; thus, to view this topic through the lens of the Google Search interface is to see a wide-ranging and contentious debate narrowed to a handful of perspectives, none of them from scholarly work published in the past five decades. But perhaps I am not seeking a comprehensive understanding of a theoretical issue; perhaps I simply want to find the nearest library? The first page of results for "library" winnows the more than 1.6 billion possible choices down to seven library branches near me, plus news results about libraries, the Library of Congress home page, and, near the bottom, the Wikipedia entry for "library." As the scholar and interface designer Liese Zahabi describes, this is almost unrecognizable compared to early search:

> Now you can simply type in your question or term, in plain language, and get specific results. Users of databases were once forced to use modified operators, to add and exclude terms, to pair and separate words—to have an understanding of how the information was organized. Now we can simply type in a question or phrase, and if the results aren't what we want, the whole system works well enough and fast enough, that we just keep rephrasing our question until we get the expected or desired result.[38]

Often even this rephrasing is unnecessary, as "users typically use very few search terms when seeking information in a search engine."[39] I did not type "find a library near me"; "library" sufficed. Nor did I have to navigate to Google.com or type the full word: when I typed "l-i-b" in my browser's URL bar, the drop-down menu provided "library" in the autocomplete suggestions.

Nor does a user need to expend much mental effort on studying the results. In the early days of Yahoo and Alta Vista, results were categorized. Google's default category is "All." It gives few visual markers to distinguish what type of website or information each result represents; any link might be a blogger's unresearched opinion or a journalistic piece by an established news institution. Visual distinctions between paid and unpaid content have diminished.[40] With so little context, users are primed to trust Google's ranked results rather than evaluate for themselves the credibility of, or relationships among, the sources of information listed. The implications are serious. A 2018 Pew study found that American adults are frequently incapable of distinguishing between news that is factual and news that is opinion-based.[41] Misinformation is a pervasive issue, particularly as Google has repeatedly refused to edit its search engine results. Despite protest from the Anti-Defamation League, Google initially did not intervene when the first page of search results for the term "Jew" included anti-Semitic Holocaust-denial websites.[42] Noble has shown that misinformation about racial violence gleaned from online search contributed to Dylan Roof's 2015 shooting of African American churchgoers in Charleston, South Carolina. PageRank's purportedly neutral ranking imposes a hierarchy, conferring the top results with authority that may be unwarranted. "Search," as Noble writes, "does not merely present pages but structures knowledge."[43]

The forms Google imposes on information through the simplicity of its interface and hierarchical display of results condition users to take a passive role in information consumption. As I have been describing, the pervasiveness of computational mediation creates the perception that information is too vast to be understood or represented in its totality. Problematically, the black box ethos prevents users from establishing a middle ground between understanding all information online and finding a single relevant result. Systems are referred to as *black boxes* when their workings cannot be understood by their users (as opposed to the cloud iconography, indicating that users *need* not understand the details of operation). A system may be a black box for proprietary reasons. In the case of PageRank, the precise workings of the algorithms are a trade secret. Complexity may also be a contributing factor. Facebook, for instance, analyzes nearly 100,000 factors to assemble each user's news feed.[44] Often those who design algorithms find it difficult to predict or understand their operations. When Netflix held a competition in 2006 to improve its recommendation engine, the company never integrated the winning algorithm into its system: "The trouble with the winning [algorithm] . . . was that while everyone could see that it was doing a better job, nobody could quite explain why."[45]

What role, then, does the print book play in an information culture dominated by scale, computational mediation, and algorithmic processing? Initially, print served as a key reference point for information online. The technology that makes Google Search possible evolved from research on automating interactions with printed material—including Google co-founders Larry Page and Sergey Brin's early work on making books searchable.[46] Early web design mimicked the conventions of print. Writing in 2001, Peter Stallybrass argued that "the navigation of computers is still imagined in the visual language that was elaborated in the fifteenth century for the navigation of books: the language of the index finger and of its prosthetic form, the bookmark."[47] Comparisons of the Internet to a library were frequent, although these served to stress the relative disorder of the former. Gorman's quotation comparing the Internet to a "vandalized library" was echoed by his contemporaries (for instance, according to Ed Krol, "what we had was a library where all the books were dumped on the floor and there was no card catalogue").[48] Google has also leveraged this metaphor: if "the web is like an ever-growing library with billions of books and no central filing system," the Google Search index is "like the index in the back of a book."[49]

Today, comparisons between print culture and online information are much less common, largely because the structures of the book and the library seem to be a poor match for the apparent formlessness of information online. Stallybrass's index fingers and bookmarks have given way to arrow cursors and starred favorites. Google Book Search invokes rhetorical and material models of text far removed from the printed materials that underpin the collection. By 2012, Google had scanned more than 20 million books.[50] In its early stages, the project, originally known as Google Print, was characterized as a "comprehensive, searchable, virtual card catalog of all books in all languages."[51]

According to Google, these were not books themselves: each "book" in the system was more akin to metadata about that book. The rationale for this comparison was legal as well as morphological. A card catalogue indexes a collection without storing its content. Google framed its project in this way to argue that it was not making copyright-protected material freely available. As Mary Sue Coleman, the president of the University of Michigan, one of Google's early partners in Google Book Search, put it: "I see no difference between an online snippet, a card catalog, or my standing at Borders and thumbing through a book to see if it interests me, if it contains the information I need, or if it doesn't really suit me."[52] We *read* print books (at least, once we have determined whether they suit us); we *search* Google Books. This distinction was emphasized with the 2005 rebranding of Google Print as Google Book Search.

Formally, too, Google Book Search reveals more differences than correspondences between print books and their digital representations. As Melissa Chalmers and Paul N. Edwards describe, "in Google's conversion system, the hard-won features of bound books—the very things that made them convenient, efficient, and durable media for so long—were treated as bugs rather than features. Google routinely excluded materials from scanning due to size or condition. These included very large or small books as well as books with tight bindings, tipped-in photographs and illustrations, foldout maps, or uncatalogued material. . . . Very old, brittle, or otherwise fragile books were also excluded." With such limitations, Google abandoned its initial goal of digitizing each of the 32 million books catalogued in WorldCat. The digital books they did produce became palimpsest objects, "comprised of a series of page images, a file containing the book's text, and associated metadata."[53] Google Book Search represents the datafication of print culture: "Google's interests lie beyond the human-readable book towards leveraging the data with the corpus to expand their algorithmic and advertising potential, offering a service unrecognisable within traditional models of the book trade."[54]

The movement away from conceptualizing the Internet with reference to print culture involves, in part, the desire to more fully embrace the unique qualities of digital media. It is also driven by an awareness of formal differences. As I have described, the scale of digital data so far surpasses that of the print-based information economy that, when references to the scale of print culture surface, they do so as a point of departure. As Thomas Vogler writes of the conceptual poet Kenneth Goldsmith's *No. III 2.7.93–10.20–96,* a six-hundred-page compendium of text combed from the Internet over a period of several years, "the effect is to emphasize the futility and inadequacy of the 'book' as antiquated 'container' for a rampant textuality that includes . . . all the verbal litter strewn across the World Wide Web. The form of the book, as permanent container of information, is the wrong vehicle for such a proliferation of the ephemeral in chaotic and uncatalogable form."[55]

Another difference is the relative fixity of print compared to the flux of information online. Information online is added to constantly; it also decays as webpages and databases are neglected, absences signaled by broken links and 404 messages. If the Internet is an archive in any sense, it is, as Wolfgang Ernst argues, "a dynamic archive, the essence of which is permanent updating."[56] Even the scanned books in Google Book Search are in a state of flux: "Google routinely updates and replaces scanned content after running it through improved error detection and image quality algorithms," which "erase a finger in the margins of a scan, restore a missing page, or deliver a once-buried quote in search results."[57] Additionally, the regime of personalization means that any user's understanding of the Internet is confined to the individualized perspective of their "filter bubble."[58] In the case of Google Search, which has personalized search since 2009, the list of results may be determined by a user's past searches, physical location, or many other factors. Search, writes Frank Pasquale, "give[s] each of us a perfect little world of our own, a world tailored so exquisitely to our individual interests and preferences that it is different from the world as seen by anyone else."[59] This world is malleable and constantly changing: the more data tracked via searches and clicks, the more the parameters change.

Print is not a perfectly static medium, nor is the print book a perfectly static platform. Books may be annotated, destroyed, or reissued with errata. The history of the book bears out how readers and writers have altered their books' content, from everyday practices such as adding marginalia to avant-garde cutup techniques. Yet the form of the book as a bound and closed object, its pages sewn or glued into a codex form, affords fixity and finitude. As Johanna Drucker notes, books "are static artifacts in the conventional sense with the finite sequence of their pages fixed into the binding" prior to their existence in the performance of reading.[60] The fixity of the book is particularly salient compared to the perceived ephemerality of digital texts. Jacques Derrida, for instance, describes "writing with ink (on skin, wood, or paper)" as "less ethereal or liquid, less wavering in its characters, and also less labile, than electronic writing," and N. Katherine Hayles opposes the "flickering signifiers" of text read on a computer screen to "the fixity of print."[61]

Additionally, the print book imposes immediately apprehensible forms onto its content. This hypermediation sharply contrasts the decontextualizing mediation of online information. A book is an organizational system that structures the content it stores. While books are limited in storage capacity by the size of print and paper, the fact that they are bound conveys a sense of comprehensiveness and completeness. A collection of individual units combined into an ordered whole, the book functions as a material metaphor signifying completion, perfection, or totality.[62] Otlet invoked this aspect of the book, likening all of the documents in his bibliographic systems to "chapters and paragraphs of a single universal book."[63] As I discussed in chapter 1, a book's physical dimensions serve as a cognitive map for the reader, imparting perceptual metadata about its organization. Books make legible not only their content but also the sets of relationships among their parts. The figure of the open book represents the material and conceptual opposite of the black box of digital information culture.

Chapter 2 FROM BIG DATA TO BIG BOOKS: SEARCHING FOR MEANING IN MATTHEW MCINTOSH'S *THEMYSTERY.DOC*

What, then, of the novel—the genre that has been shaped by its long association with the book? As I have discussed, theorists of the novel have defined the genre by its scale as well as its heterogenous assemblage structure. As such, it has a formal similarity to the Internet. Matthew McIntosh's *theMystery.doc* explores this similarity, demonstrating that print novels, while unable to contain a tiny fraction of the quantity of information online, make information meaningful by subjecting it to the constraints of narrative, page, and book. *theMystery. doc* is a maximalist novel: at nearly 1,700 pages, it has been described by reviewers as "weighty," a "brick," and part of a "door-stopping tendency" in American maximalist literature.[64] With its considerable bulk—it weighs roughly four and a half pounds—*theMystery.doc* asserts the thingness of the book. At the same time, the title employs the naming conventions of a word-processing file. The work juxtaposes the unwieldiness of the book with a digital storage format that takes up little room on a hard drive, highlighting how differently textual media may embody large amounts of information. The scale of the Internet may be vertiginous, but rarely do we feel the imposing presence of information online as directly as we do with this print novel.

theMystery.doc also generates its sense of scale formally: its fragmented, polyvocal text gestures toward more than the sum of its parts. Novelistic printed prose is only one of its many genres and media. McIntosh incorporates photographs, movie stills, transcriptions of sound recordings and telephone calls, logs of online chats, content copied from webpages, a hand-drawn picture, and much more. There are blank pages, pages filled with asterisks (representing everything from snow to static to elementary particles), and more than four hundred images. Narratively, the novel weaves together disparate strands—fictional, autobiographical, and archival. The central fictional story is that of Daniel, who awakens one morning with no memory of who he is. He knows only that he has spent the past eleven years writing a massive novel and that no one knows what it is about. (The title stems from Daniel's discovery of a computer file named "themystery.doc"; when he opens it, the document is blank). While Daniel is a fictionalized stand-in for McIntosh, *theMystery.doc* also presents apparently nonfictional retellings of seminal moments from McIntosh's life (apparently, because the ontological status of these elements is difficult to determine). This McIntosh persona (referred to as "Matthew," "Matt," "M," and "Matthew McIntosh") is a narrativized construct, but these sections include photographs and transcriptions of documents depicting the personal events narrated (the death of the author's father; the funeral of his baby niece; McIntosh's first novel, *Well*). The third stratum of the novel is the archival layer, consisting of the text of documents relating to historical events—most centrally, a website documenting the disappearance of a missing woman named Kim Forbes and a transcript of a telephone call between an emergency operator and a woman who died in the World Trade Center during the September 11 attacks.

At the heart of these narratives lies the search for meaning. The titular mystery refers diegetically to the subject of Daniel's novel. No one is able to state what his novel is about, and many of *theMystery.doc*'s metafictional moments concern the various author-personas' quests to write massive novels capable of making sense of the world. There are other diegetic mysteries, from whether Daniel will recover from his amnesia to whether the chat transcripts are records of conversations with other humans or with chatbots. More broadly, the title evokes the religious denotation of *mystery* (the novel documents visions, sermons, and proselytizing); the question of America's future as a nation; and individuals' attempts to find meaning in a world in which death and suffering are random, unavoidable, and devastating.

theMystery.doc exemplifies an ongoing trend in the contemporary novel: employing novelistic scale and narrative complexity to comment on information culture. Analyzing this trend, Stefano Ercolino defines the maximalist novel as "an aesthetically hybrid genre of the contemporary novel" among whose signature characteristics are length, encyclopedism, and dissonant chorality.[65] (At 1,664 pages, *theMystery.doc* is longer by a significant margin than Ercolino's two longest examples: David Foster Wallace's *Infinite Jest* at 1,079 pages and Roberto Bolaño's *2666* at 1,105 pages.) Ercolino's is one of a number of critical accounts of the contemporary novel's information density. John Johnston argues that novelists such as Thomas Pynchon and Don DeLillo "engage [the reader] with various kinds of [information] multiplicity, both by registering the world as a multiplicity and by articulating new multiplicities through novel orderings and narrativizations of heterogeneous kinds of information."[66] Ercolino builds on Tom LeClair's concept of the "systems novel" and Frederick R. Karl's "Mega-Novel." In a similar vein, Hayles reads contemporary fiction via the history of information theory and cybernetics.[67] Although these critical works are grounded in diverse models of complexity, and although they disagree on whether the novels they study tend toward totalizing ordering systems or chaotic multiplicities, all view the large-scale, post-1970 American novel as a genre that grapples with complexity, density, and information. The scale trend may also be present in the contemporary novel writ large: according to one study, the average length of books has increased by about eighty pages per year since 1999.[68]

This aspect of *theMystery.doc,* then, is not new; we find it in the work of Wallace, Pynchon, and DeLillo, among others. (I will have more to say on the whiteness, maleness, and Americanness of this canon below.) What is new—and what distinguishes the meta*medial* novel's emphasis on its embodiment in print from previous meta*fictional* novelistic work on information scale—is the exploration of how the organizing form of the print book creates an alternative conceptualization of information to that created by online search interfaces and other processes of computational mediation. McIntosh uses online search as a figure for the quest for answers rather than their successful retrieval. The link between the search for meaning and online search is made before the reader has opened the book: the cover features the title, "theMystery.doc," and, beneath this, the words "a novel" in a search bar (a white rectangle with a magnifying glass icon). The cursor still shows after the *l* in "novel," as if someone has just typed this phrase. Within the novel, there are several series of search results. The most thematically significant are the results for the query "Kim Forbes," which are interspersed with the story of a fatal car accident and a list of deaths from natural disasters.

Kimberly Oswald Forbes was a real woman who disappeared in Oregon in 2004. Her disappearance remains a cold case. I discovered who Forbes was by performing my own Google search. *theMystery.doc* encourages readers to search online to try to determine which elements are fictional (or at least fictionalized). By folding search results into the narrative, McIntosh demonstrates the centrality of online search to the production and consumption of contemporary fiction. The content of the results for "Kim Forbes," moreover, contrasts the randomness of information encountered online with the meaningfulness of information situated within a narrative and a print book. Here are some of the first results McIntosh includes:

> > "forbes " <kim.forbes@——————————— > wrote in message news: 1121353283.698009.10770@g14g2000cwa.google groups.com...> We are doing more and more ...

> > .kim forbes. Male, 35 nassau Bahamas Last Login: 12/2/2007 1:37 PM ... About kim forbes. im fun loving person witty humor i love intelligent conversation i ...

> 11, 4–8, Angie Yohe, 1995. 4–8 Melanie Long, 1998. 13, 4–4, Kristen Miner, 2006. 4–4, Kim Forbes, 2006. 15, 4–2 12, Krista Oyler, 1995 ...

Chapter 2   > Bush Seeks Nuclear Disclosure from Kim—Forbes.com[69]

I have attempted to preserve McIntosh's formatting, as these "linguistic marks left by the network ecology" demonstrate how much of the text encountered online is directed to nonhuman readers.[70] As Dworkin describes, such marks become awkward "when moved to the substrate of the printed codex page": "Randomly generated strings of . . . web-address file names appear unremarkable and remain all but unread by online viewers when they announce a clickable link . . . , but they become ungainly when read aloud or typed out on the page."[71] Several of the Kim Forbes examples are difficult to parse for this reason. Their narrative significance is also initially unclear. Who is Kim Forbes, and what do these various search results—some of which, like the *Forbes. com* article, are obviously not about a person named "Kim Forbes"—reveal? As the reader progresses through the book, however, the story of Forbes's disappearance arises from the noise:

> P.S. Remember "Kim Forbes"—she simply disappeared—no one has heard from her in almost ten years ... Photo Sharing and Video Hosting at Photobucket...

> Hood River—KIMBERLY FORBES (Endangered Missing) (452, ellipses in original)

The section culminates in pages of messages left on a website (still accessible at the time of this writing) dedicated to Forbes's disappearance.

By situating these online traces of "Kim Forbes"—the missing woman as well as the search string—within the context of a novel and the space of a book, McIntosh makes several points about the relationships among online information, the print novel, and the human search for meaning. First, it suggests existential loss. While Forbes dominates the section that bears her name, we never learn much about her beyond the facts of her disappearance. The search for "Kim Forbes" online, like the real-life search for the missing woman, bears few fruitful results. Hers is only one of many stories of death and loss in the book, from the teenager who dies in the car crash, whose suffering is recounted in visceral detail, to the lists of unnamed people who perished in natural disasters.

Second, the Kim Forbes example shows that meaning in the novel is accretive and is thus the antithesis of the randomness of information found online. Zahabi warns of the danger of conflating search results with a sophisticated understanding of a complex question: "because the question is complicated, and because there are so many different 'correct' choices, a search engine (currently at least) cannot just spit out the one right answer. The complexity of the human condition, the complicated nature of knowledge, the extensive connections to be made within culture—these are features, not bugs to be fixed."[72] In the case of Kim Forbes, the scale of the Internet is more a liability than an asset. The Internet's ability to network concerned citizens and share information has not helped solve this mystery. It does, as the support messages show, provide a platform for offering consolation, but even this community formation is minimized in the noise of the many unrelated search results. The purposeful arrangement of these search fragments in the novel, in contrast, builds associations between them and the surrounding narratives as well as gradually bringing the story of Forbes's disappearance and its emotional consequences into focus through its sequencing. One chapter in the Kim Forbes section is titled "Basic Facts"; facts, however, become meaningful information only insofar as they are situated in relation to other events and to larger questions.

theMystery.doc uses the print book to make sense of an increasingly mediated information culture. As a book-object that signifies through its mass, *theMystery.doc* is a book first and a novel second. When M speaks with an unnamed interlocutor about his work in progress, both note the manuscript's size: "It sounds like a really, really, really huge scope [*sic*]," says the questioner, to which M responds, "It is scope itself" (991). When the interlocutor asks whether readers will be daunted by the size, M answers, "I'm not trying to get anybody to read it" (997).[73] McIntosh again raises the issue of his work's potential unreadability—and indicates the modernist roots of his metamedial investigation of scale—when a character fails to read past the first page of Joyce's *Ulysses*. McIntosh reproduces the first paragraph of Joyce's book, including its iconic typography: the capital *S* of the opening word "Stately" extends into the first two lines, mimicking the large initial *S* of the novel's first American edition, designed by Ernst Reichl. (As Jessica Pressman notes, the opening page of each narrative in *Only Revolutions* also begins with an enlarged initial letter, again recalling the typography of the 1934 *Ulysses* edition; this move in *Only Revolutions* similarly "locates the contemporary novel in a genealogy that reaches back to Joyce and to modernist poetics.")[74] Whereas the postmodern novel is an archive of literary intertextuality, the metamedial novel is also a typographic archive. The opening of *Ulysses* is reproduced twice more in *theMystery .doc,* as the narrator remarks, "I couldn't concentrate. I kept going back to the beginning"; eventually, Joyce's text breaks into "ahh, forget it," as the narrator gives up (1520, 1521). At the level of the page, McIntosh's text visually merges with Joyce's; as a large, bright-blue book printed with white letters, *theMystery.doc* also mimics Joyce's book design.

theMystery.doc's argument is finally that one's understanding of the world—however limited, however incapable of staving off suffering or death—emerges through a process of contextualization, enacted in this novel through the form of the codex. Meaning is created through the juxtapositions that occur sequentially (as in the progression of "Kim Forbes" search results) and across page spreads. In one such sequence, photographs of an elderly couple dancing at a senior center face images of a glamorous couple dancing in a black-and-white film. The effect would be lost in a Kindle edition, where each page fills the screen in isolation; it would also be lost if we were to scroll through the pages rather than turn them, as the latter interaction makes the figures on the page spin as though dancing. Likewise, the novel's blank pages signify, becoming empty computer screens, representations of the emptiness of death, or symbols of future histories still to be written. The narrative concludes with twenty-three pages, blank except for pagination and the running title, followed by another three that are completely blank sheets. "The novel's vocation today," writes Daniel Punday, lies in its "ability to represent the absent, potential, or unrealized."[75] This ability lies not only in narrative but also in the metamedial mobilization of the print book's conventions.

theMystery.doc privileges connections forged by humans—those made between disparate pieces of information as well as interpersonal connections. Romantic and familial relationships are archived through transcripts, narratives, and photographs. Near the novel's end, McIntosh affirms relationality by referencing his real-life wife, Erin: on one side of a page spread is the dedication "*for Erin,*" while the other side features a blurry photograph of a smiling woman (presumably the dedicatee) (1586–87). "Mrs. Matthew McIntosh" is also credited with the book's design (untitled copyright page). As these acknowledgments of his wife indicate, a romantic relationship was central to the making of *theMystery.doc* as a novel and as a book. Relationships, the work implies, are similarly constitutive of reading: the point is not to solve the mystery definitively but to build connections among the various fragments in this manuscript. These fragments, many collected from the digital trails left by people's movements online, cohere to form something more meaningful. In this way, moments from McIntosh's life are connected to the lives of his family members, to the suffering of strangers, to recent American history, and to the question of how narrative might merge fiction and reality.

This literary emphasis on making meaning through formal accretion and organization arises in response to the scale of digital information and the stakes of books as information media. *theMystery.doc* repositions the human at the center of meaning making, refuting computational search's negation of human decision making. Moreover, *theMystery.doc* links this humanistic information processing to the form of the book. In its bulk, layout, and spatiality, the book makes the organizational connections among its parts apprehensible—the opposite of the decontextualized information retrieval that is central to the aesthetics of digital mediation and computational search. When M's wife receives an email about a visit to the cemetery on what would have been their niece's first birthday, McIntosh leaves in the advertising boilerplate at the email's end: "Don't just search. Find. Check out the new MSN Search!" (1041). This text recalls the biblical imperative "seek, and ye shall find." The point of *theMystery.doc* is not so much to find answers as to grapple with the process of seeking them out. Ultimate meaning may only ever be immanent, but the embodiment of large-scale and heterogenous texts and data in a book allows the novel to foreground how actively we must seek when we seek for meaning.

WOR(L)D PROCESSING: THE COMPRESSION AESTHETICS OF MARK Z. DANIELEWSKI'S *ONLY REVOLUTIONS*

Mark Z. Danielewski's *Only Revolutions* shapes the infoscapes of American and world history into a book. It recounts the parallel stories of Sam and Hailey, teenagers who fall in love as they travel across the country. Their journey is also symbolic, as they transition from the apparent freedom of solitude to finding fulfillment in their ties to one another. Merging the road trip with the epic, the narratives are written in lyric verse and set against the backdrop of history. Sam's and Hailey's stories occur at a remove in time from one another, running from November 22, 1863, to November 22, 1963, and from November 22, 1963, to January 19, 2063, respectively. Their lives progress in tandem with events from the American Civil War, through the Kennedy assassination, and into the future of the United States. The duo become archetypes, encompassing the stories of tragic lovers from Bonnie and Clyde and Tristan and Iseult to John and Jacqueline Kennedy.

This narrative scope recalls the encyclopedic impulse of the maximalist novel, an impulse to which Danielewski is no stranger. His *House of Leaves* (2000) is a novelistic compendium of genres and media, and the first five volumes of his *The Familiar* series (2015–17) run to more than 4,000 pages.[76] In contrast, *Only Revolutions* is modest in length: 360 pages, the number of degrees in one complete revolution. Moreover, its text is governed by strict constraints. Each chapter consists of eight pages, each of which has exactly ninety words of narrative. The lexicon is carefully delimited, with the end pages listing words that do not appear in the text. Danielewski establishes the historical contexts with a feature he calls the *chronomosaics*: fragmented lists, printed in the gutters of each page, that parallel the narratives typographically and thematically. Monotonously recording death counts, snippets from famous quotations, and other cryptic references to historical events, the chronomosaics evoke the scale of history but are highly abridged.

Whereas *theMystery.doc* represents information scale through its size and voracious inclusivity, *Only Revolutions* represents it by juxtaposing the challenging density of the text's information gluts with the limited spaces of the novel and the book. The perception of scale is created through the effort the text requires from the reader. *Only Revolutions* challenges conventional readability quite literally at every turn. Sam's and Hailey's narratives are printed at a 180-degree rotation from one another. Each page is a multidirectional grid of texts: the first page of Hailey's narrative contains the last page of Sam's, printed upside down, as well as the corresponding chronomosaics. The reader must choose her own route through the book—a significant metaphor, as the book's columns of texts visually represent the lanes of a highway and the reader must turn the book as if turning a steering wheel.

Only Revolutions uses the technique of compression to encode the scale of global information within the confines of the book. This

Nov 22 1963
—to screaming.
—he's gone.

Parkland Hospital.
1:00 PM.
Oak Cliff & 2 cartridges,
JD Tippit goes.

—This is it.
Lee Harvey Oswald.
—I haven't shot
anybody.

2:41 PM
Love Field.
Air Force One. LBJ &
Judge Sarah T Hughes.

—defend, protect and.
—OK, let'S get this
plane back to.

John W McCormack's
security squad.

—Serious but not.
—a giant Cedar.
Broadway.
Half-mast.

—Would you come
with US?

—That is all I can do.
I ask for your help.

—a tragedy for all
of US.

Dow down 21.16.

Samsara! Samarra!
Grand!
I can walk away
from anything.
Everyone loves
the Dream but I kill it.

Atlas Mountain Cedars gush
over me: —*Up Boogaloo!*
I leap free this spring.
On fire. How my hair curls.
I'll destroy the World.
That's all. Big ruin all
around. With a wiggle.
With a waggle. A spin.
Allmighty sixteen and freeeeeee.
Rebounding on bare feet.

Trembling Aspens are pretty here:
—*You've nothing to lose. Go ahead.*
Have it all.

Tamarack Pines sway scared.
Appalled. Allso pretty. Perfumed.
Why don't I have any shoes?

I could never walk away from you.
but who cares for it? O Hailey no,
Everyone betrays the Dream
and allways our hushes returning anew.
petals & stems bending and lush,
so long it keeps turning with flurry & gush,
I'll destroy no World
what your joy so dangerously resumes.
And I, your sentry of ice, shall allways protect
By you, this World has everything left to lose.
Garland of Spring's Sacred Bloom.
And all her patience now assumes.
Solitude. Hailey's bare feet.

—0 mdl 0 mi—
Head. Cap. 80 MPH.
Greer & brakes. Roy Kellerman. Tague.
—his shot.
—We are hit.
Adeline Kennedy,
John B Connally.
Grassy Knoll. Rufus Youngblood.
Texas School Depository. 411,
Elm Street. Dealey Plaza.
Motorcade. 1235 PM.
Van Thanh Cao & 31 others.
South Vietnam Revolutionary Council,
Democrats. Book Mart & 200,000.
Love Field. 1137 AM. Grassroots
—we are going to live with Them.
& a toss.
Pooralh & Alouette crash, 5 go.
United Arab Republic, North Korea
Margaret Dico.
8:45 AM. Fort Worth Dallas.
Nov 22 1963

FIGURE 1. The first page of Hailey's narrative in *Only Revolutions*. From *Only Revolutions: A Novel* by Mark Z. Danielewski, copyright © 2006 by Mark Z. Danielewski. Used by permission of Pantheon Books, an imprint of the Knopf Doubleday Publishing Group, a division of Penguin Random House LLC. All rights reserved.

project is signaled by the algorithmic logic that underpins the novel. Scholarship on *Only Revolutions* has largely concentrated on excavating its patterns. The constraints that govern its textual organization turn the book's bibliographic codes into processural elements, resulting in "the book as a computer: a calculating machine that generates algorithms and geometrizes the plane and the space for writing and reading."[77] Academic and lay readers have declared that *Only Revolutions* defies close reading—that the massive effort needed to unpack its patterns and allusions to history outweighs the insights gained. My analysis focuses on the chronomosaics, as they most clearly demonstrate this difficulty. The chronomosaics are marginal in every sense. They are excluded from the audiobook version and the publisher's guidelines for navigating the print text, and they line the pages' gutters. The narratives' themes—ill-fated lovers, violence, revolution, and so on— are amplified by the historical references, yet this principle is easily grasped without reading most of the individual entries, the majority of which are unintelligible without research. As Hayles argues, "a complex exploration of the connections between the narratives and entries" would be "a nearly impossible (and certainly tedious) task"; for her, the purpose of the chronomosaics is to "gesture toward a vast ocean of data."[78] Careful readers of *House of Leaves* are rewarded with secret codes that shed light on the novel's central mysteries. With *Only Revolutions,* the act of decoding, rather than uncovering hidden layers, restores the full content of the original layer. To read the text is to decompress the world within it.

Broadly, *to compress* is to constrain in space; technically, *to compress* is to determine "the minimal physical resources . . . that are necessary and sufficient to faithfully encode or represent information."[79] An object, whether "a number of a bitstream or a dynamical system," is compressible if it "can be expressed in a different way in fewer bits."[80] Take the example of audio compression: "to make an MP3, a program called an encoder takes a .wav file . . . and compares it to a mathematical model of the gaps in human hearing. . . . . It discards the parts of the audio signal that are unlikely to be audible. It then reorganizes repetitive and redundant data in the recording, and produces a much smaller file."[81] As a literary technique, compression allows a text to imply scale beyond its actual size. For example, I. A. Richards describes T. S. Eliot's use of poetic allusion as "a technical device for compression": "'The Waste Land' is the equivalent in content to an epic. Without this device twelve books would have been needed."[82] (Note that Richards conceptualizes the cultural scope of the work via the physical size of books.) This description of literature as compression, in the general sense of a condensation or distillation of the external world into language, has surfaced periodically in accounts of twentieth- and twenty-first-century literature.[83] *Only Revolutions*'s model of literary compression is more technical, informed by the meaning of information compression.

Describing his desire for an aesthetic of brevity, Danielewski has described "beauty" as "provid[ing] an enormous compression of data."[84] Such compression is most obvious in the chronomosaics. While the chronomosaics are akin to the reference genre of the chronicle, they allude only briefly to the historical events they index. Instead, their "brief, elliptical entries can be considered a form of code, containing hidden messages and resonances."[85] Take the first six entries for "May 23 1984," from the midway point of Hailey's narrative (page 180):

> Kristine Holderied.
> Second Hand Smoke.
> Colleen Renee Brockman goes.
> Saudi Arabia's 400
> Stinger missiles.
> —*monstrous carbuncle on.*[86]

This list cannot immediately be understood. It must be deciphered. The full events to which these lines allude are as follows: Holderied became the first woman to graduate at the top of the class at the U.S. Naval Academy on May 24, 1984. On May 22, the surgeon general announced a definite link between secondhand smoke and lung disease. Brockman was one of dozens of women murdered by the Green River Killer; her body was discovered on May 26. On May 29, the United States announced that it would be sending four hundred antiaircraft Stinger missiles to Saudi Arabia. The italicized quotation is from England's Prince Charles, describing an extension to London's National Gallery as "a monstrous carbuncle on the face of a much-loved and elegant friend." "An enormous compression of data," indeed.

In *Only Revolutions,* readers must actively confront gaps in the text's readability, considering how individual fragments of data are transformed into information as they are placed into larger contexts of meaning. There are thousands of entries in the chronomosaics, linked only loosely through the accidental associations of history. The effort and attention required for a reader to unpack each fragment is Herculean. As I described in chapter 1, readers are likely to skim lists, especially when they are not ordered according to narrative logic. Yet the chronomosaic can be decompressed. The "text" of the chronomosaic is not only the printed words but also the content latent in its references. The novel's full complexity would be realized by a reading that deciphered every entry and charted each of the complex correspondences that link Sam's and Hailey's narratives to one another as well as to the chronomosaics (and parallel moments in the chronomosaics to one another). Although an individual reader may be unlikely to trace all of these correspondences, they are nonetheless present. They exist as the informational potential of the novel, compressed into the text printed on the page.

That such close reading work is possible is indicated by the chronomosaics' gradual progression in complexity. Sam's story begins in 1863; all entries on the corresponding chronomosaic reference the American Civil War. Some are immediately familiar ("Ulysses S Grant"), creating a framework for deciphering the less obvious fragments (e.g., "—*mostest*" derives from a quotation attributed to Confederate lieutenant general Nathan Bedford Forrest: "Get there firstest, with the mostest"). As the opening page trains us to read the chronomosaics, it also reveals their connections to the narratives. For instance, Sam's third utterance, "Contraband!," a term used during the Civil War to describe fugitives from slavery, corresponds to the two lines printed on the same axis in the chronomosaic, "—*Abolition of slavery, / confiscation of property.*" Sam is surrounded by "Bald Eagles" whose cry of "—*Reveille Rebel!*" stresses the chronomosaic's emphasis on the Civil War. Set against this backdrop, Sam's and Hailey's personal revolutions for freedom are figures for (and are figured by) the Civil War. This connection is strengthened by the fact that Sam and Hailey refer to themselves as "US." The complexity increases when Sam's opening page is read in conjunction with Hailey's. In Hailey's first chronomosaic, every entry relates to the Kennedy assassination, another turbulent moment in American history. "Atlas Mountain Cedars" speak to Hailey, whose chronomosaic records "—*a giant Cedar,*" referencing a speech by the Senate chaplain: "The President of the Republic goes down like a giant cedar."[87] This line establishes "go" as a euphemism for "die." A reader who skips this chronomosaic will miss the reasoning behind the novel's use of "go"—a word that occurs more than 3,000 times—to mean death. Conflating death and movement, "go" is *Only Revolutions's* defining trope. Read in conjunction with the chronomosaics, Sam's and Hailey's road trip becomes a literal death-drive.

Prompting the reader to uncover these connections, *Only Revolutions* confronts the decontextualization that results from online search's algorithmic mediation of information. Danielewski proposes instead that information, however great in magnitude, must be carefully contextualized and closely examined. The novel flirts with the idea of totalizing systems: its dominant circular motifs, like Sam's and Hailey's progression to sexual and legal union, represent the desire for comprehensive wholeness. As Hayles observes, *Only Revolutions* "puts information excess into tension with an elaborate set of constraints."[88] Ruptures constantly threaten any attempt to achieve completeness. The novel oscillates between unity and division (captured in the prominent portmanteau word "allone," combining "all one" and "alone"). Sam and Hailey initially chafe at social restraints, and the novel's antagonist, the CREEP, represents the social orders and conventions from which the duo try to escape.[89] Yet the novel also suggests that these powers may be changed by "socially embedded individuality, even if it exists in the smallest possible community of two: as long as self and other enter a relationship, their individualities enable them to affect the world."[90] Sam's and Hailey's quest for equality—an equality that manifests typographically as they approach each other in equal-sized font at the book's center—is layered onto references to key legislative moments in American Civil Rights history, such as *Brown v. Board of Education* and the Civil Rights Act.[91] Subjectivity and agency exist on a spectrum of interconnectedness, and the meaning of any individual act or event is immanent in its relationality.

Only Revolutions insists that data are like an ego: only meaningful when set in a relational context, even if this is one of many possible contexts. Sam and Hailey reach their full expression through their relation to one another and to their historical contexts. Similarly, each fragment in the chronomosaic is meaningful insofar as readers link it to its historical context and to the narratives—just as Google search results become meaningful within broader social contexts. Indeed, literary reading can be coextensive with Internet search, and each can illuminate the other.[92] If the chronomosaics seem unreadable in practical terms, imagine how much more daunting the task of deciphering them would have been before the web. The chronomosaic entries are so decontextualized that there exists no obvious method of tracking down their references without digital search resources such as Google. *Only Revolutions* is poised on the boundary between digital and predigital knowledge archives. While the highly abbreviated nature of the chronomosaic entries requires readers to integrate online search into their interpretive methods, the historical research Danielewski undertook to write the chronomosaics was decidedly un-digital. The bulk of his research was done in libraries, consulting newspaper archives, microfilm, and historical print sources—the kinds of media his modernist predecessors had access to. Danielewski also created his own mini-archive, a collection of index cards recording handwritten notes about the historical vignettes that would become the chronomosaic entries.[93] These individual items were then recontextualized through their incorporation into the chronomosaics. Perhaps this immersion in print and other predigital media explains why *Only Revolutions* differs from digital information systems in making its gaps visible, requiring the reader to consider how meaning making occurs.

Only Revolutions's bookishness contributes to its information contextualization. Of all Danielewski's works, this one has most escaped its print boundaries: the author narrates the audiobook, and the e-book includes music and hyperlinked annotations. Yet the print novel's bookishness is impossible to overlook, with its multidirectional text layouts and multicolored inks. As I have noted previously, one effect of metamediality in the novel is to defamiliarize the conventional attributes of books and reading (especially novelistic reading) that are often taken for granted. The novels I have been studying constitute a series of test cases that either work against these defaults or assess their consequences. In the case of *Only Revolutions,* one insight regards how the book's fixity imposes an organizational frame on its content. In the terms of systems theory, a book is a closed system. A closed system prohibits energy, information, or physical matter from moving across the system's boundaries. The Internet is an open system, in that it can be continuously updated; so is Danielewski's collection of index cards. By contrast, the informational content in a printed book is fixed by default. As I have been arguing in this chapter, the fixity of print on the page, and of pages arranged in the codex, allows readers to mentally map the textual content, establishing sets of relationships among the books' parts. Reading is a form of embodied cognition. As such, "the material substrate of paper provides physical, tactile, spatiotemporally fixed cues to the length of text" because "the reader can see as well as tactilely feel the spatial extension and physical dimensions of the text."[94] *Only Revolutions* demonstrates that the cognitive processes that form associations between individual parts are tied to the spatial relationality of the printed page and the codex.

Only Revolutions invokes the algorithmic as a model for narrative, but it foregrounds the mediations of the book and the agency of the reader. The novel demands a cyborg reading, wedding human with search interface; its narrative connections, however, prioritize the humanistic work of interpretation. *Only Revolutions* also harnesses another kind of cyborg reading: that of networked communities. Sam and Hailey relate to one another, to the United States, to the entire world, and also to the novel's readers. Their use of the word "US" "refers at once to the exclusivity of two lovers preoccupied with each other while the world whizzes by, the national collective of America, and a transnational community of readers stretching across time and place."[95] Danielewski has stated that the ideal reader of *Only Revolutions* might be a computer, but a full excavation of the work's potential text—its latent relationships, awaiting recognition—is already underway via his online readership.[96] He has hosted web forums devoted to interpreting his works since the early 2000s. At the time of this writing, the *Only Revolutions* forum has nearly 8,000 posts. In 2018, he launched online book clubs, where readers convened on Facebook to analyze *The Familiar, House of Leaves,* and *Only Revolutions.* The dedication of *Only Revolutions*—"You were there"—invites the reader into the text. It registers how essential readers are to the work's interpretation.[97]

Collective interpretations of literature are not new; as Lisa Nakamura writes, "books have always been a means of social networking."[98] Yet crowdsourced interpretation, where a novel becomes the catalyst and context for an online network of readers, stands as an alternative model of reading in the digital age. The network becomes an interpretive framework. As we have seen in this chapter, the dominant ethos of Internet search, and of algorithmic culture more broadly, is that technology must mediate between humans and information—that the scale of information is beyond the individual's ability to manage or navigate. Networked reading serves as a way to approach the scale and complexity of a text (and of the cultural archives on which a text draws) without ceding interpretive agency. One human mind is insufficient to the task of excavating a collection of information as large or dense as the chronomosaics, but Danielewski's use of socially networked reading illustrates how networks of human minds generate processing power greater than the sum of their parts. Like Amazon's Mechanical Turk Web service, which is grounded in the rationale that sufficiently large groups of humans routinely outperform computers at certain tasks, a social network of readers derives its impact from its own scale—from its amplification of the individual into the multitude.

FRAMING THE NOVEL

theMystery.doc and *Only Revolutions* demonstrate an intervention the novel is well situated to make in the twenty-first century: to heighten awareness of the interface dynamics endemic to information culture. Of course, one might object to this argument on the grounds that these two texts are not unambiguously novels. *Only Revolutions*'s narratives are written in verse; *theMystery.doc* incorporates found documents and memoir. As I have been arguing, however, part of what is at stake in the uneasy genre status of such works is a self-conscious reevaluation of the limits, and the potential contributions, of the novel in an information culture distinguished by the obfuscations of scale. Metamedial novels challenge conventional genre designations because they explore how novelistic narrative may have informational as well as literary properties.

A number of experimental writing practices have engaged with the relationships among literature, information, and mediation, but the majority have fallen under the rubric of poetry. As Lori Emerson has discussed, poets including Emily Dickinson, midcentury typewriter poets, and authors of early electronic concrete poetry have addressed the function of technological interfaces.[99] In the twenty-first century, the literary examination of information culture has frequently been the task of conceptual literature, a designation that encompasses a variety of experimental practices, including "appropriation, piracy, flarf, identity theft, [and] sampling."[100] Conceptual writing is largely understood to be a matter of poetics.[101] Kenneth Goldsmith's *The Weather* (2005) consists of transcriptions of weather reports; Judith Goldman's "dicktée" (2001) lists every word from *Moby Dick* that begins with the prefix "un-." Goldsmith reads the rise and proliferation of conceptual literature as a direct response to the quantity of texts available online, the accessibility of these texts via broadband and mobile media, and the ease with which computers allow writers to copy and paste. "Faced with an unprecedented amount of available text," he argues, "the problem is not needing to write more of it; instead, we must learn to negotiate the vast quantity that exists."[102]

Like *Only Revolutions* and *theMystery.doc* (and *Ulysses* before them), conceptual poetry transforms the quotidian language of information into the literary. Literariness is a notoriously elusive quality; any definition must be expansive enough to cover substantial differences in genre and historical context and specific enough to be critically incisive. The most common formulation hinges on an opposition between what is literary and what is informational. Walter Benjamin insisted that the "essential quality" of a literary work is decidedly "not statement or the imparting of information." For Ludwig Wittgenstein, "a poem . . . is not used in the language game of giving information."[103] These definitions have proven to be robust, especially in scholarship on the interconnections between literary culture and information culture.[104] Because the novel has been routinely described as an inherently informational genre, it has a formal and theoretical investment in assessing the boundaries between the informational and the literary. Metamedial novels do this by producing tension between the inclusive flexibility of the designation *novel* on the one hand and the containing and framing of elements within the aesthetic context of the novel and the physical space of the book on the other.

Previously in this book, I have considered what J. Paul Hunter calls "the promiscuous inclusivity of the novel," a quality he links to the genre's association with print.[105] The novel has been a formal testing ground, defined by length or the presence of prose rather than by consistently distinctive formal features. I began this chapter with reference to *Macroanalysis*. For Jockers, the issue with traditional literary studies boils down to exemplarity:

> Big data render [close reading] totally inappropriate as a method of studying literary history. This is not to imply that scholars have been wholly unsuccessful in employing close reading to the study of literary history. A careful reader, such as Ian Watt, argues that elements leading to the rise of the novel could be detected and teased out of the writings of Defoe, Richardson, and Fielding. Watt's study is magnificent . . . but he has observed only a small space. What are we to do with the other three to five thousand works of fiction published in the eighteenth century?[106]

Jockers's solution is to seek lexical patterns across thousands of novels using computational analysis. It is true enough that the theory of the novel has historically proceeded through the analysis of canonical examples, a concession to practicality as well to academic valuation. Yet, as I have been arguing, part of the work of the novel is to explore the potentialities of literary form. While Jockers's work is compelling, there is something profoundly disconcerting in his proposal that the novel is best understood by identifying the largest patterns across the greatest numbers of texts. After all, per Bakhtin, the novel is "ever questing, ever examining itself and subjecting its established forms to review."[107] Scholars of the novel must theorize not only the broad patterns novels exhibit but also how the novel, as a genre, functions as a series of individual cases, each of which may develop its own formal approaches.

It is difficult to imagine, for instance, that a computer program trained on eighteenth-, nineteenth-, and twentieth-century novels—or indeed, other twenty-first-century novels—would identify either *Only Revolutions* or *theMystery.doc* as novels. True, some reviewers have seen the tagline "a novel" to be flippant, ironic, or a marketing tool in these two cases. Several scholars have argued that *Only Revolutions* is not a novel due to the fact that it is composed of verse and lists.[108] Danielewski, however, has indicated his desire to locate his work within the tradition of the novel. He views the novel as a flexible and shifting form, a "compost heap" that "provides a place for that which is excluded" from other literary forms."[109]

theMystery.doc frames its genre designation in terms of information classification and digital search. "So like *pretend* that *I* work for LexisNexis and I'm gonna write an abstract for the search engine," says an unnamed interlocutor (992). How would M classify it? M is unable to answer definitively; he cannot even answer the apparently straightforward question "in the Dewey Decimal System would it be under fiction or non-fiction" (1000). His increasingly exasperated interlocutor sighs, "You're like: *Trust me. All—all Library of Congress systems, all Dewey Decimal Systems, will be eradicated by this book. It will break—it will break two hundred years of the classification of books . . .* Yeah, but. Yeah, but I mean . . . Somebody's gonna put it in a box that you're not comfortable with, dude, because like somebody's gonna want to check it out" (1000, ellipses in original). Formally and ontologically, *theMystery.doc* strains at the boundaries of its genre designation. Thematically, however, this story of a struggle with identity and death, of a *Künstlerroman,* and of the connection between the individual and the nation ticks the boxes of the American novel.

The designation *novel* conditions the reader to view a work's discrete elements as part of a cohesive collection with internal structure (if not, following Lukács, as a totality). Although information-dense novels represent scale through their scope and disorienting juxtaposition of fragments, they also highlight the importance of selection. As Goldsmith argues, when digital resources make the amassing of great quantities of text a trivial task, the significant factor is what the author chooses: "If it's a matter of simply cutting and pasting the entire Internet into a Microsoft Word document, then what becomes important is what you—the author—decides [*sic*] to choose. Success lies in knowing what to include and—more important—what to leave out."[110] (Recall the end pages of *Only Revolutions,* which list words excluded from the book.) The literary effect of novels such as *Only Revolutions* and *theMystery.doc* stems from their selections, alterations, and combinations of extant informational writing set within narrative, generic, and medial frameworks that create a recognizably aesthetic object, an object whose internal structures and connections the reader can tease out.

Take the example of Karen Reimer's *Legendary, Lexical, Loquacious Love* (1996), which merges distant reading, conceptual writing, and genre fiction. Reimer selected a romance novel and reordered all of its words alphabetically. The first and second pages consist of the word/letter "A"; page 3 branches out to "abandon," "abducted," "aboard," and "Aborigines." The text, marketed as "An Adult Romance for the Post Structuralist Woman," resembles Julia Flanders's description of the products of "early text analysis tools": "Their basic outputs were essentially the text in a dismembered state—concordances, word frequency lists, collocations."[111] *Legendary, Lexical, Loquacious Love* both is and is not the original text. Imposing the order of the alphabet over the order of narrative fundamentally alters the story. The patterns one encounters in Reimer's text result from the chance associations of spelling. Yet the text demonstrates that narrative and affect may inhere in even a seemingly randomized dataset. Flanders writes of early text-analysis outputs that "the remoteness of these representations . . . from 'the text' as a readerly artifact underscored the sense that a great interpretive distance would have to be traversed in order to bring this data back to the realm of literary meaning."[112] *Legendary, Lexical, Loquacious Love,* however, like the chronomosaics of *Only Revolutions,* reveals that "the text . . . in a data-like form" is still located within "the realm of literary meaning." That early list of *a* words—"abandon," "abducted," "aboard," "Aborigines"—suggests a voyage to Australia, a suspicion that is confirmed on page 36.

Moreover, as Reimer describes, the result of her rearrangement of the source text is a novel whose emotional qualities are produced by the interactions between narrative and data: "I wondered how the love story would exist without a narrative structure/plot. I used the alphabet—an arbitrary, non-hierarchical ordering convention—for its objective unemotional character, which places it at odds with the subjective emotional character of romance novels. I wanted to see whether, if I put a romance novel's cover on it, it would be possible to read an alphabetical list of words as a love story. I think it is."[113] *Legendary, Lexical, Loquacious Love* lays bare the conventions and quirks of the romance genre, repeating words like "corset" (nine times) and "breast/s" (sixty). There is often an oddly affective quality to the insistent repetition, as when the text reiterates the heroine's name, culminating in instances with question marks and exclamation points: "'Anastasia?' 'Anastasia?' 'Anastasia' 'Anastasia!' 'Anastasia!' 'Anastasia!'"[114] Reimer's deformance of the original text becomes its own aesthetic (and narrative) entity.

As the framework of the novel conditions readers to discover patterns in Reimer's text, her choice to publish the work as a print book—a mass-market paperback—conditions them to identify it as a novel. The text is printed on thin paper, and the spine is delicate. Its cover displays standard pulp-romance fare: a woman, dress falling from her shoulders, swoons in the arms of a dark-haired man. As Dworkin notes, works of conceptual literature have frequently been published first as digital texts and then as printed texts, usurping the book to bestow a culturally recognizable literary form on their content. While Dworkin argues that this practice is less prevalent in recent conceptual literature, the book continues to provide writers with a vehicle for framing found-digital materials as literature. In an analysis of Robert Fitterman's "No, Wait. Yep. Definitely Still Hate Myself" (2014), which collects expressions of loneliness from social media posts, Dworkin argues that the work "benefits from being a printed book, with more intensive protocols of reading than the screen," and that "the book, as a unit, establishes the frame against which its irony can emerge."[115] We recognize *Only Revolutions* as a novel in part because it draws our attention to the fact that it is printed in a book. When M in *theMystery.doc* is asked what he is writing, he settles on "a big *book* about America" (989, emphasis added). We recognize *theMystery.doc* as a novel in part because so many maximalist novels have been big books about America. In metamedial novels, the book refracts the scale of Internet-era information culture into paperspace.

Interfaces dominate the contemporary information landscape. They mediate the ways in which we seek information, communicate, and consume text. In one respect, today's culture of computational mediation is a continuation of the methods and guiding principles established in the early twentieth century: the atomization of information, the automation of search and retrieval, the transition from the book to more modular media. For these reasons, modernist writers such as Joyce remain literary touchstones as today's novelists engage with information and its management. Twenty-first-century mediation, however, also involves computational and algorithmic processes that intercede between humans and the information they desire. Tools such as Google Search are a practical way to navigate the staggering amount of information available online. Yet digital interfaces do not eliminate or reduce the complexity of large-scale information. Instead, they mask it, preventing users from perceiving or confronting what that scale consists of and how it might be otherwise imagined or managed.

Metamedial novels offer another perspective, foregrounding the book's presentation and arrangement of information. They remind us that any collection of text, from a novel to a list of search results, is necessarily limited. Internet interfaces elide, glossing over the complexity that they manage; novels amalgamate or compress, making visible the processes of mediation that constitute their literary practices. These novels ask us to confront the scale of digital information culture, even as they confront the limited scale of their own medium. We need not abandon digital search to take seriously the alternative visions of information management these novels create. Interrogating how we read, search for, and interpret textual information, *theMystery.doc* and *Only Revolutions* make a literary intervention into critical algorithm studies.[116] These texts ask, How else might we imagine, conceptualize, or interact with information at scale? What would a hypermediate search interface look like—a search interface that emphasizes its selection methods and attempts to contextualize its results?

One of the most important lessons from media history is that media are rarely neutral. Phonography was shaped by its listening audiences, becoming "defined amid and as part of late nineteenth-century and early twentieth-century tensions surrounding the role of women and other 'others' in U. S. society," and demonstrating "that media and their publics coevolve."[117] The practice of basing sound design on the acoustic preferences of "expert listeners"—a practice that formed the basis of sound design from early telecommunications through the development of the MP3—assumed that the ears of white middle-class men could represent all listeners.[118] Google's insistence that its algorithms are unbiased obscures the fact that these algorithms have historically been developed by a narrow demographic. Consequently, search results reproduce stereotypical depictions of people of color, women, and other marginalized groups.[119]

As I have noted, this chapter focuses on two novels written by white American men. While the novel of information scale is not a strictly American phenomenon, maximalist and encyclopedic literature has been disproportionately written by American authors and, more specifically, by white American men—Pynchon, DeLillo, and so on. This concentration is perhaps unsurprising when we consider how central the hegemonic perspective of the "white male subject position, that unmarked position that humanism idealizes as universal" (as Kathleen Fitzpatrick puts it), has been to the history of information classification and how it continues to influence the development of search algorithms.[120] Yet *Only Revolutions*'s and *theMystery.doc*'s argument that the embedding of fragments of data in larger social contexts is necessary to fully understand them provides, if not a way out of that hegemonic perspective, at least a mirror held up to it.[121]

In this respect, these metamedial engagements with information culture contest the decontextualization that has been championed by conceptual writers. In a chapter titled "Why Appropriation?" Goldsmith writes that "the digital environment has completely changed the literary playing field" by creating "a time when the amount of language is rising exponentially, combined with greater access to the tools with which to manage, manipulate, and massage those words." Given these conditions, he concludes, "appropriation is bound to become just another tool in the writers' toolbox."[122] Goldsmith completely sidesteps the questions of whether and to what extent appropriation as literary practice overlaps with the politics of cultural appropriation—an issue that came to a head with his controversial "The Body of Michael Brown" (2015), which appropriates the autopsy report of Michael Brown, an African American eighteen-year-old shot by a police officer in 2014. In a cogent critique of this brand of conceptualism, which holds that any text may be recontextualized by any author without regard for the social or political implications, Cathy Park Hong writes that Goldsmith's "PoMo for Dummies 'no history because of the internet' declarations became absurdly irrelevant when black men were dying at the hands of cops," and argues that "the era of Conceptual Poetry's ahistorical nihilism is over."[123]

While *Only Revolutions* and *theMystery.doc* share conceptual writing's interest in remixing informational texts into literature, they insist, contra Goldsmith, that information is always already embedded in the contexts of the personal, the local, and the social. Approaches to information studies drawing on feminist critique and critical race theory have "highlight[ed] the importance of positionality—the situatedness of knowledge."[124] As metamedial novels interrogate the material conditions of reading and of information organization in the age of computational mediation, they draw parallels between text as it is physically situated in a book, information as it is organizationally situated in a narrative, data as they are semantically situated within larger systems of meaning, and information as it relates to the social world that produces and consumes it.

CHAPTER 3

Haptic Storage

Disembodied Information, Textual Materiality,

and the Representation of the Subject

We like to feel ... that other hands have been before us, smoothing the leather until the corners are rounded and blunt, turning the pages until they are yellow and dog's-eared. We like to summon before us the ghosts of those old readers who have read their Arcadia *from this very copy.*

—Virginia Woolf, "The Countess of Pembroke's Arcadia"

This reading film unrolls beneath a ... strong magnifying glass five or six inches long set in a reading slit, ... and the reader is rid at last of the cumbersome book, the inconvenience of holding its bulk, turning its pages, keeping them clean, jiggling his weary eyes back and forth in the awkward pursuit of words from the upper left hand corner to the lower right, all over the vast confusing reading surface of a columned page.

—Bob Brown, The Readies

A NEW FORM OF THE BOOK

In 1911, the *Miami Metropolis* published Thomas Edison's predictions for life in the year 2011. One concerned the future of the book. Edison declared that books could meet the growing demands of modern information management if they could be reengineered to store greater quantities of text: "Books of the coming century will all be printed leaves of nickel, so light to hold that the reader can enjoy a small library in a single volume. A book two inches thick will contain forty thousand pages, the equivalent of a hundred volumes."[1] The foundation of this prediction—the assumption that the chief limitation of books is the amount of information they can store in a given space—continues to drive comparisons between print and digital media a century later. Edison's solution did not gain much traction (although, as I mentioned in chapter 1, the *Encyclopaedia Britannica* published that same year was printed on very thin India paper). A different solution had been put forth half a decade earlier by Paul Otlet (co-founder of the Mundaneum) and Robert Goldschmidt. Like Edison, Otlet and Goldschmidt presumed that print books were poorly scaled for information management; unlike Edison, they proposed to replace books altogether. Otlet and Goldschmidt proclaimed microform "a new form of the book" that could produce crucial space savings by replacing books with microphotographic images of their pages.[2] In the future they anticipated, readers would access microform content via portable viewers, magnified screens set into reading stations, or text projected onto walls. Instead of being books "so light to hold," as Edison had suggested, these texts would not typically be held by readers at all.

Thus far in this study, I have examined a key aspect of modern information scale and its management: how information scientists have developed search apparatuses that help users locate specific pieces of data but that conceal the underlying organizational systems—a situation contested by novelists, who have leveraged the print book's hypermediacy to challenge this obscuration of form. In these cases, the problem posed by scale has primarily been one of too much content. But media themselves could also contribute to the problem of scale by occupying too much space. In this chapter and the next, I examine the view that the book is ineffective for information management because of its size and other physical properties. At issue is information's embodiment: the ways in which data become meaningful through the forms they take as they are stored in media. In the early decades of the twentieth century, the perceived need to mitigate information proliferation by shrinking storage media spurred claims that micro-form would replace the book. Because microform necessitates the use of interface screens to access the stored content, its adoption drove the belief that information's material form was inconsequential. The rhetorical position that microform could usefully deemphasize or alto-gether negate information's embodiment ran counter to the metamedial investment in the book's textual materiality that I have been arguing was an important strategy in the modernist novel.

These changing ideas about information, media, and embodiment had implications for another aspect of information management during the modernist period. As Otlet and La Fontaine were considering the prospect of archiving all of the world's information, many of their contemporaries were exploring the scale of personal information—the representation of individual subjects as collections of data. These information professionals debated what kinds of data were necessary for this task as well as how much would suffice to represent a life. Personal data were increasingly available: people generated substantial paper trails simply by going about their daily business (paying bills, buying tickets for mass transportation, filling out forms, and so on), and governments systematically monitored their populaces, turning national bodies into data sets. In the information culture of the early twentieth century, there was thus a double negation of embodiment: with the promised future of microform, information was seen as less intrinsically embodied (a message separable from its medium), even as people became subjects that could be quantified as information—their identities interpellated by records, files, and documents rather than constituted through their lived existence.

Both of these shifts called the relevance of the novel's archival proj-ect into question. The rise of microform threatened to imperil the print book: the novel's paradigmatic medium and the one that has shaped how novelists conceive of their genre's informational dimensions. The belief that an individual could be represented as information in turn decentered the novel's role as the genre of modern subjectivity. Much criticism on modernist narrative form has discussed how novelists developed techniques such as stream-of-consciousness writing to model and theorize subjectivity. These representations of the subjectivity of the modern individual contrasted with the positivist and ostensibly objective representations of the subject as an accumulation of data. How, then, might novelistic representations of individual subjects imitate or critique the notion of the self as constituted through information? And how did perceptions about information's instantiation in media work in tandem with, or against, conceptualizations of selfhood as abstracted—that is, not tied to life in a physical body?

I read Virginia Woolf's *Orlando* as an attempt to resolve the discrepancy between these competing versions of represented identity. In my view, her attentiveness to the embodiment of the print book highlights how books may preserve aspects of the material world not only by representing them (via narrative descriptions or collections of biographical data) but also by accumulating objects and other physical traces as they are handled by readers. Books are tactile objects that accumulate forensic information in this way; their hypermediacy makes their own embodiment readily apprehensible, and they have been closely connected to human bodies throughout the history of their design, production, and use. Thus, they have emblematized a medial metaphysics of presence. Books therefore offered novelists such as Woolf a means of opposing the ideology of information's disembodiment, especially as this concerned the abstraction of people into information. *Orlando* is a fantastic tale, but is also a *roman à clef,* Woolf's attempt to capture her estranged lover. The writer reflects on the limits of representation, concluding that neither a novelistic narrative nor a vast quantity of biographical information alone can represent an individual life but that attention to the storage medium's material qualities may complement both approaches. *Orlando* demonstrates one way in which metamedial novels have intervened in discussions about modern information mediation: by exploring the affective and archival registers of the book's materiality.

MANAGING MATERIALITY: THE FISKEOSCOPE,
BOB BROWN'S READIES MACHINE, AND THE
REVOLUTIONARY POTENTIAL OF MICROFORM

We do not customarily list microform among the media that sparked the imagination of modernist authors. *Microform* is the general term for the medium of microscale texts—"for any information storage and communication medium containing images too small to be read with the naked eye."[3] Microform includes specific formats such as microfilm (on reels) and microfiche (on cards). Largely obsolescent today, microform lacks the aesthetic appeal of similarly outdated media such as vinyl records and Polaroid cameras. It has tended to be the subject of either apathy or antipathy.[4] Scholars are less and less likely to exhume microfilm reels from their storage boxes as the medium is increasingly eclipsed by digital technology. As Leah Price puts it, microfilm is "now too old to be sexy but too new to be quaint."[5] Perhaps for these reasons, it has been, in the assessment of Jonathan Auerbach and Lisa Gitelman, "neglected by cultural historians as well as cinema scholars."[6] With regard to the modernist period, the early twentieth century was a fallow time for microform, caught between the newness of microphotography's invention in 1839 and microform's much more widespread and visible application following the Second World War. Yet the first decades of the twentieth century brought a critical shift in the medium's use, from storing images to reproducing texts.

John Benjamin Dancer produced the first microphotograph in 1839. Within two decades, *Photographic News* anticipated that "the whole archives of a nation might be packed away in a small snuff box [using microphotography]. Had the art been known in the time of Omar, the destruction of the Alexandrian Library would not have been a total loss."[7] Although the very first microphotograph was a reduced document, microphotography in its early decades was primarily used to reproduce pictures (for example, in jewelry with microphotographic portraits embedded inside the baubles). Reproducing images rather than words,

Haptic Storage

microform was a medium more closely aligned with pictorial photography than with writing. This changed at the turn of the twentieth century as advocates began to discuss the benefits of replacing print texts with microform, particularly for large-scale information management.[8]

Otlet and Goldschmidt argued that using microphotographic reproductions of books would save storage space, lower costs, facilitate scholarly access to research materials, and aid in the preservation of valuable manuscripts. Books were problematically bulky, "heavy to handle and tak[ing] up a relatively large amount of space." The "microphotographic book," in contrast, would "do away with the [print book's] inconveniences . . . [instead] produc[ing] books that are: 1) less heavy and smaller; 2) uniform in size; 3) on a permanent material; 4) moderate in price; 5) easy to preserve; 6) easy to consult; and 7) continuously produced."[9] If books were transferred to microform using Otlet and Goldschmidt's microfiche card format, a single ten-drawer card catalogue would be able to hold the equivalent of 268 meters' worth of library shelves—close to 19,000 volumes.[10] The driving force was the perceived mismatch between the scale of books and the scale of information. Books were paradoxically too large to be efficient uses of space and not large enough to store sufficient quantities of information. If books were replaced by microform, Otlet and Goldschmidt claimed, "all of Human Thought could be held in a few hundred catalogue drawers."[11]

In the 1920s, the use of microform became widespread. George McCarthy's Checkograph, patented in 1925, brought it into the management of banking records. That same year, the inventor Emmanuel Goldberg demonstrated "microfilm reduction equivalent to putting the entire text of the Bible fifty times over on one square inch of film."[12] As a result of these and other developments, by the end of the decade microform was being used by businesses; by criminologists, for handwriting analysis; by individual scholars, who made microscopic copies of their research materials; and by organizations such as the League of Nations and the Library of Congress. This momentum explains why, although microphotographic processes had been invented in the 1830s, Eugene Power could describe them as a "very new technique" a century later.[13] "It will not be long," wrote the librarian Edward A. Henry in 1932, "before our larger research libraries will have thousands of volumes of books on film on their shelves and perhaps large reading rooms equipped with projectors for their use."[14]

Because a microformed page is literally a negative of a print page, the medium served as the book's conceptual negative: portable where the book was heavy, compact where the book was bulky, uniform where the book was irregular, new where the book was old. David Thorburn and Henry Jenkins have demonstrated that "utopian and dystopian visions" have been "a notable feature of earlier moments of cultural and technological transition—the advent of the printing press, the development of still photography, the mass media of the nineteenth century, the telegraph, the telephone, the motion picture, broadcast television."[15] We might add microform to this list. In 1935, Harry Miller Lydenberg, the director of the New York Public Library, asked, "Is the library world witnessing a transition from the volume form of the book to roll form [of microfilm], a reversal of the change that came nearly two thousand years ago when book makers turned from the roll form to the codex?"[16] Lydenberg refrained from answering his own question, but others in the burgeoning field of information science would have answered with a resounding "Yes." By the mid-1930s, we encounter statements such as "the application of the camera to the production of literature ranks next to that of the printing press [in importance]" and "there is taking place in the techniques of record and communication a series of changes more revolutionary in their possible impact upon culture than the invention of printing."[17] If the discourse surrounding the potential impact of microform can be dismissed in hindsight as "the naïve and impractical imaginings of overenthusiastic pioneers," to quote one historian of the medium, it is nonetheless a significant, and critically neglected, element of the information culture of the early twentieth century.[18]

While such rhetoric tended to attribute microform's revolutionary potential to its space savings, microform's most significant potential for the revolution of information culture came through its use of mediating interface screens. Microform texts are, by definition, unreadable without the assistance of a reading apparatus. These interfaces displaced the stored text from the reader's view, concealing the underlying medium in a manner akin to the concealments of modern information mediation I have described in previous chapters. In so doing, microform's interfaces reconfigured the physical experience of reading. In the epigraph with which I began this chapter, Woolf imagines a community of readers constructed through the marks they have left on the pages of a book. With microform, readers no longer handled the medium's surface while reading. They avoided leaving fingerprints that could smudge the microphotographic image. The metal surfaces of the viewing platforms lacked paper's tendency to accumulate traces, and many reading platforms were freestanding, designed so that the reader could sit back and read at a distance. Discouraging touch, microform tended to negate the haptic modes of preservation associated with the print book.

From the vantage point of the twenty-first century, microform's fate is clear: after the Second World War, formats such as microfilm and microfiche flourished in specialized information management contexts, like libraries, banks, and government departments; with the advent of digitization, they began to lose their hold in these areas. Today microform is associated with information storage rather than literature. During the modernist period, however, its future was open, and many microform innovators proposed that microphotographic texts should be used for private pleasure reading. One microform reading apparatus intended at least in part for the personal consumption of literature was invented by Bradley M. Fiske, an American naval admiral. When Fiske received his patent in 1922—the first patent issued for a reading machine in the United States—he already had a successful record as an inventor. (The *New Yorker* called Fiske "one of the most notable naval inventors of all time.")[19] The device, which would be called variously the "Fiskeoscope," the "Fiske-o-scope," and the "Fiske Reading Machine," was a handheld reader designed to be used with one eye.

The Fiskeoscope quickly garnered national attention: it was featured in publications including the *Library Journal, Popular Mechanics,* and *Scientific American.* The *New York Times* described Fiske's machine as "a tiny affair" consisting "of a light frame of aluminum, which carries a strip of paper containing reading matter whose characters have been reduced by photo-engraving to a size about one-one-hundredth

Haptic Storage

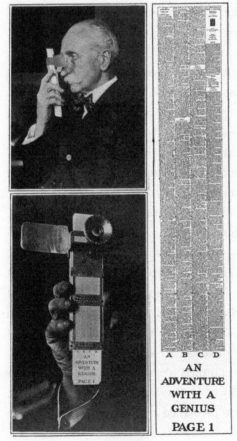

FIGURE 2. Admiral Fiske and the Fiskeoscope, *Scientific American* 6 (June 1922).

of the size of ordinary typewritten characters. A lens surmounts the frame, and through this the characters are magnified."[20] According to the *Library Journal,* the device was "held on the nose like a lorgnette" as "a thumb screw revolves each page into view."[21] The Fiskeoscope's characteristics varied across a number of patents for different iterations of the device, but the most publicized version had the reading matter reproduced on long rectangular strips, which Fiske called pages. The Fiskeoscope generated considerable interest. In addition to receiving attention in the *New York Times,* publications such as the *Library Journal* continued to discuss the device more than a decade after it was patented. *Popular Mechanics* and *Scientific American* reproduced the microphotographic pages used by the machine within their own pages, as if literalizing its ability to rupture the dominance of print.

What was modern about this new mode of reading, as the news coverage of the Fiskeoscope makes clear, was its unobtrusive physicality. The Fiskeoscope represented "simplicity and compactness personified."[22] We might be reading an advertisement for the Kindle: "the whole device weighs little more than five ounces and is intended to be used with one hand. It can be carried in the coat pocket without causing a bulge. The instrument is . . . only a quarter of an inch thick."[23] Compare Otlet and Goldschmidt's recommendation, published a few years after Fiske received his patent, for a reading machine "so small that it is pocket size."[24] Fiske claimed that the device would improve eyesight as well as drastically reduce publishing costs, but its most immediate benefit was its "enormous condensation in the bulk of reading matter."[25] *Popular Mechanics* wrote that the Fiskeoscope would be useful for reducing "bulky reference works," and the publishers of the New York telephone directory considered producing a microform phonebook to be read on the Fiskeoscope.[26] As these accounts demonstrate, the materiality of books—their "bulk," their weight, the space they occupy—was construed as a burden that microform would relieve.

Fiske believed that these qualities would allow the Fiskeoscope to gain traction in the realm of pleasure reading. "A person who likes books, but who must move so often he cannot collect them," according to the inventor, "might carry a 50 or 100 volume library in a cigar box." "In the near future, when you pack your bag for a vacation trip," declared a reporter in a similar vein, "you will be able to take along with you a library with as much reading as the Encyclopedia Britannica, packing it in an ordinary suitcase, and have enough room to spare for an extra shirt or two."[27] Most telling is Fiske's assertion that "a 'best seller' of 100,000 words . . . costing $1.50 can be produced for four cents per copy."[28] Significantly, this description suggests that popular literature (a "best seller") would be read on his device. He did not want his machine to be limited to viewing reference materials but hoped that the Fiskeoscope would turn microform into a literary medium. To this end, he publicized the device by reducing Mark Twain's best-selling travel narrative *The Innocents Abroad* to thirteen pages.[29] Space savings made microform competitive with books for storing information. With portable microform readers, the medium could also compete with books for storing narrative.

Fiske was one of a number of people to advocate for personal, portable microform readers. Even the literary avant-garde had interest in the subject. The period's most vivid engagement with microform's literary implications arrived in the form of Bob Brown's "Readies" machine. A 1941 newspaper article described Brown as "the man usually credited with having the first idea of reading by machine and who invented the most radical version of it."[30] Brown's experiments with the mediation and materiality of text had begun on paper: among his varied career efforts, he had published visual poetry, which incorporated handwritten text as well as sketches. His most unusual literary experiment, however, was designing a machine that automatically scrolled microscale texts before the reader. Like Fiske, he believed that the principles of microform should be applied to literature. Also like Fiske, he extolled microform's ability to do away with the unwieldy physicality of books.

In 1930, Brown published a manifesto in the journal *transition* calling for a "revolution of the word," one brought about by his automatic reading machine.[31] Originally conceived in 1916, the device sounds something like a cross between microfilm and tickertape: "Books could be printed in miniature type on a roll of tape and reeled off by an electric motor beneath an oblong magnifying glass."[32] Inspired by cinema—"the Talkies"—Brown named the text written for his machine "Readies." As the *Chicago Tribune* described Brown's machine, "it is to resemble a typewriter in shape, but will be much smaller. The printing will be done on a ribbon of tough impressionable material and will unravel, electrically, much as a typewriter ribbon progresses. . . . In the center of the machine, . . . the actual reading will be done through a magnifying glass. The speed at which the ribbon will unravel can be regulated to the velocity desired by pressing buttons."[33] The text would be microscopic and printed on tape or film (Brown's specifications varied over the machine's development), but this ribbon would spool text in a single stream, like tickertape. In addition to automatically moving text across a reading screen, Brown wanted his device to be portable, "a simple reading machine which [he could] carry or move around. . . . A machine as handy as a portable phonograph, typewriter or radio" (*The Readies,* 28).

In the two books Brown published on the subject in 1930 and 1931, his writing seems to echo both Futurism and Taylorism. "Modern word-conveyors are needed now," he proclaimed; "reading will have to be done by machine; *microscopic type* . . . brought up to life size before the reader's birdlike eye, saving white space, making words more moving, out-distancing the flatulent winded ones and bringing the moment brightly to us" (*Readies,* 13, emphasis added). Despite the performative prose and borderline-facetious tone of his writings about the Readies machine, Brown put significant effort into the device's design and production, underscoring his serious interest. A newspaper article describes him as having "made two working models of his machine," and his anthology of Readies texts, *Readies for Bob Brown's Machine,*

FIGURE 3. Bob Brown's Readies Machine. Artwork by Elffin Burrill, in *Readies for Bob Brown's Machine,* ed. Craig Saper and Eric White (Edinburgh: Edinburgh University Press, 2019), based on a photograph glued into copies of *Readies for Bob Brown's Machine* (Roving Eye Press, 1931).

includes a photograph of a prototype.[34] In one of a number of letters that he wrote to Gertrude Stein discussing the machine, Brown mentions that "two or three publishers are considering making a machine or printing the [Readies anthology] book."[35] He harbored hopes that the machine would be manufactured in time for the anthology to be "printed in miniature type for the first roll to be used on the machine," and he corresponded with a patent lawyer as well as potential manufacturers in Detroit and Moscow.[36] Ultimately, Brown was unable to solve the design problems posed by the Readies machine. Could the text be printed onto tape rather than photographically reproduced? Could the words be projected onto the reading screen rather than read through a magnifying device? Ruing his lack of capital, he wrote to Stein, sending her information about the machine along with the hint that "maybe some of your millionaire friends would actually buy one [of my machines]."[37]

Brown never produced a commercial model of the Readies machine, and the project thus failed to catalyze the reading revolution that he had predicted. If the machine was a material failure, however, it was nonetheless a conceptual success: it inspired modernism's literary elite to consider what it might be like to dispense with the print book, reading not the page but the screen. As Craig Saper describes, Brown was "a central character in the avant-garde arts, both among the expatriate writers, publishers, and artists in France and among the Imagist poets." "His friends and mentors included cultural figures from Marcel Duchamp and Man Ray to H. L. Mencken and O. Henry to Nancy Cunard and Gertrude Stein to William Carlos Williams and Mina Loy."[38] Leveraging his connections to these modernist networks, Brown involved a number of the literary avant-garde's leading figures in the Readies project. Stein, Williams, Marinetti, Ezra Pound, Eugene Jolas, and three dozen other writers contributed "Readie" texts, which he published in *Readies for Bob Brown's Machine*. As Michael North notes, many of the authors who contributed to the anthology "sent previously published works or squibs like Williams's 'Readie Poem,' which didn't even pretend to take the new invention very seriously." Still, "a few prominent writers, including James T. Farrell and Kay Boyle, 'translated' earlier works for the readie machine, and at least a few of these seemed to be carried along by Brown's faith that his was the most promising front in 'the Revolution of the Word.'"[39] Brown attributed the project's origins in part to modernist writing. "I had to think of the Readies machine," he wrote, "because I read Gertrude Stein and tape-tickers" ("Appendix," 206). In a literary take on technological determinism, he also claimed that his device's machinic efficiency would inevitably make language more dynamic: to produce texts readable at high speeds, writers would have "to eliminate all sluggish, uninteresting writing."[40] Brown's call for the "revolution of the word" was both literal and figurative: his revolution in reading technology would spark a revolution in experimental writing as his machine revolved words across a screen.

During the past decade, Brown has become the subject of increased critical attention, coinciding with literary criticism's growing interest in media studies and book history.[41] What previous work on the Readies has missed is the extent to which his writings about the machine are in keeping with early-twentieth-century discourse about microform's revolutionary implications. While scholars have discussed how technologies such as the cinema and the tachistoscope influenced Brown and have considered how the Readies machine anticipated aspects of electronic reading, few have mentioned his machine's indebtedness to microform.[42] He did not adopt the conventions of microform formats wholesale: his choice of a single, scrolling line of text diverged from most microform documents, which typically reproduced print pages, and he critiqued this aspect of the Fiskeoscope (*Readies,* 34). Yet microscopic type—"printing done microscopically by the new photographic process"—was essential to the Readies (*Readies,* 28). Brown frequently compared his machine to the Fiskeoscope; in fact, he discussed his own plans for a reading machine with Fiske, in person and by mail, and told the admiral that he "treasure[d] a leaf from the New York Telephone book made for [Fiske's] machine."[43] Microform's influence was also apparent in Brown's 1931 work *Words,* which juxtaposed poems printed in standard type with poems printed in microscopic type. As Saper notes, "for readers to see the smudge at the bottom corner of the page as a microscopic poem, they need[ed] an external apparatus."[44] "All that is needed to modernize reading," Brown wrote, "is a little imagination and a high powered magnifying glass" (*Readies,* 31–32). His plan capitalized on the growing fascination with microscale texts in the late 1920s and 1930s, and his bombastic rhetoric about the doomed future of the book drew on the discursive modes through which microform was understood and imagined.

Chapter 3

DISEMBODIED INFORMATION VERSUS MATERIAL TEXTS: FROM MICROFORM'S MEDIAL IDEOLOGY TO MODERNIST TYPOGRAPHY

The manner in which microform readers' interfaces mediated between human readers and stored content gave rise to the perception that information could become disembodied, transcending its medium. Brown's discussion of the Readies machine illustrates this. If, he reasoned, books could be replaced with a medium that needed to be neither touched nor held, then texts would become immaterial "moving type spectacle[s]" (*Readies,* 27). To read a book was to be confined to the realm of "the clumsy hand," whereas to read words on a screen was to rely only on "the natural celerity of the eye and the mind."[45] The very register of the haptic was anathema to Brown. Even holding a book was exhausting: "I like to read big books like Mencken's 'American Language,' but I find it physically impossible to keep such a weighty tome in a pleasant reading position for long" ("Appendix," 205). He dramatized the benefits of his machine by inviting readers to acknowledge the physical effort involved in reading his codex-bound *Readies for Bob Brown's Machine:* "Now shift this tiresome book in your hand, prop up your eyelids with match sticks and move your eyes wearily back and forth over another three thousand lines or so" (161). In contrast, he imagined, his machine would allow readers to lie completely motionless, effortlessly consuming Readies as the words were transmitted to their minds as if by magic. As Paul Stephens describes, "Brown's idealized reader becomes one with information."[46] Brown, of course, ignored the extent to which sight and thought are embodied processes. He ignored, too, the reality that microform texts and their reading devices are often awkward to use precisely because of the ways

in which they are embodied, requiring readers to laboriously manipulate zoom lenses and dials in order to bring the texts into focus.[47]
Yet he repeatedly insisted that his machine, by substituting words on
a screen for pages in a book, would radically alter the embodiment of
reading and of information itself.

The rationale behind this assertion becomes clearer with reference
to microform history. Even more than cinema—which North has
shown to be an influence on Brown's portrayal of poems as words of
light on screens—microform changed what it meant to access stored
text.[48] While much prior research on Brown has focused on the transformative effects of the machine's speed, attention to microform history
reveals the extent to which the microscopic scale of the text and the
use of the interface were equally transformative. His descriptions of
words projected directly into readers' minds demonstrates the extent to
which these interfaces underpinned a particular fantasy of disembodiment. The Readies, like microform more generally, created an *avant
la lettre* version of what Matthew Kirschenbaum has called the "medial
ideology" of electronic media, "the notion that in place of inscription,
mechanism, [and] sweat of the brow . . . there is light, reason, and
energy."[49] With microform, the physicality of stored information was
doubly deemphasized: the microform texts themselves took up far less
space, and the interface screen necessary to read them meant that the
reader no longer had to hold or touch the textual medium. Brown's
laments about the hardships of reading print books are humorously
exaggerated, but his writings reiterate the claim that "reading by
machine"—through interfaces—would fundamentally transform
the experiences of reading, phenomenologically as well as physically.

Although few microform developers made claims as extreme as
Brown's, many shared his goal of deemphasizing information's medial
embodiment. Throughout the writing about microform, the refrain
that sounds again and again is that the physicality of print media—and,
especially, of the book—is a problem best solved through its elimi-
nation. Recall Otlet and Goldschmidt's emphasis on how little space

microform texts occupy or newspaper articles about the Fiskeoscope's ability to make texts more portable. Microform did not make these texts cease to exist, but it did make them less physically intrusive than "the cumbersome book" (*Readies,* 29). (It is no surprise that one of the earliest and most persistent uses of microfilm was in espionage: documents on film were more easily secreted on the body and smuggled across borders.)[50] Brown, Fiske, Otlet, and other innovators repeatedly asserted that microform's materiality would be experienced very differently from that of the book. A microformed page is no less embodied than a printed page; they are, however, differently embodied, and the distinct postures and procedures necessary for reading microscale texts led to the belief that the information they contain could be transmitted immaterially.

The medial ideology of microform had implications for the novel. This was the case both because, as I will describe, the print novel's affordance of haptic storage worked against the microform dream of immaterial information and because many microform advocates viewed the novel as the ideal literary genre for their medium. Fiske produced at least one microform novel. While most of the texts in the Readies anthology are poems or short fiction, Brown also wanted to liberate novels from the physicality of print books, and he planned to use works such as *Adventures of Huckleberry Finn* to market the device ("Appendix," 206). The genre's default lengthiness made novels an obvious choice for microform's space-savings (as well as, theoretically, for Brown's speed reading: he argued somewhat speciously that his machine would allow readers to consume "hundred thousand word novels in ten minutes" [*Readies,* 28]). Yet he also singled out the novels of Anthony Trollope, citing their length, as the kind of writing he hoped would be destroyed by the new, more efficient writing ushered in by his machine: "Maybe no more trilogies will be written when Readies are in vogue" (*Readies,* 32). Microform had the potential to transform—or eradicate—the novel and its medial support.

Whereas microform proponents expressed a desire to make media disappear, modernist writers emphasized the materiality of literary media. Graphic innovation in advertising, along with what Jerome McGann calls the "Renaissance of Printing," made the first decades of the twentieth century a time when "practitioners of both visual arts and literature paid unprecedented attention to the specificity and formal properties of their media."[51] The dominant aesthetic was hypermediacy. While Brown wanted a "revolution of the word" divorced from the book, writers such as Marinetti wanted a "typographic revolution" that would reinvent the printed page by foregrounding its physicality.[52] Futurist and Vorticist publications highlighted the type on the page as such. Writers involved themselves in book production, with some (including Woolf and Cunard) hand-printing their own novels and others producing proto-artists' books that interrogated the book's form and cultural status.

In these cases, the book was no crystal goblet, to cite Beatrice Warde's foundational ideal of the unobtrusively "transparent" text.[53] Modernism's textual media mattered as matter. Experimental manipulation of the typographic surface and physical medium began to disrupt "the dominant Anglo-American tradition of twentieth-century typographic design, which maintains an unobtrusive clarity as its aim."[54] It also disrupted the convention that literary texts—in particular, novels—were "unmarked" and thus not notable for their layout.[55] Rather than an incidental container of information, the printed page had become materially significant. The legacy of this experimentation is important when we consider how the novel mediates and forms data: when the medium itself signifies, textual content becomes only one component of the information system.

The majority of literary typography from the modernist period comes from poetry. This was partly due to the ease of reproducing visually complex short pieces as compared to longer texts. Large commercial presses generally lacked interest in publishing experimental works, so difficult typesetting tasks tended to be left to smaller presses. Donna Rhein describes the difficulties that the Hogarth Press encountered: "*Paris* [by Hope Mirrlees] is the worst example, but the poems *The Waste Land* [by T. S. Eliot] and *Parallax* [by Nancy Cunard] and [E. M.] Forster's book *Pharos and Pharillon* with its mixture of prose, dialogue, poetry and footnotes provide other kinds of typographical horrors for the typesetter and layout designer. The time and effort involved make these productions very expensive."[56] Although the early-twentieth-century novel incorporated less typographic play than its poetic counterparts did, what did occur challenges received ideas about the genre as the transparent bearer of information. In previous chapters, I have discussed how modernist novels such as *Ulysses* and *Manhattan Transfer* make the text visible as text. Textual experimentation in fact arose quite early in the novel's history: after the title character of Samuel Richardson's *Clarissa* is raped, the text switches to a "distracting, visually confused and over-articulated layout" with text printed sideways and at odd angles, "mirror[ing] the accompanying verbal descriptions of Clarissa's distracted, confused, and inarticulate thoughts."[57] Laurence Stern's *Tristram Shandy* is another early example.

That the novel has value and significance beyond its informational content is similarly evident in the early twentieth century's fine printing tradition. Lawrence Rainey has documented how modernist special editions conflated aesthetic and commercial value, troubling the received narrative that sets modernist elitism against the interests of the commercial mass public. These editions also have consequences for a study of how the novel participates in information storage. As Rainey writes, the care with which deluxe editions were made beautiful "could lead to a paradoxical state of affairs, one in which active readers were slowly replaced by passive consumers, mere buyers who were less engaged with a book's contents and more bedazzled by its wrappings."[58] This is an ironic reversal of the theory that book and text should be transparent vessels, critical only in the information they contain. As the objectified book becomes an economic investment, the market speculator who purchases it can disregard the contents altogether. Special editions might go unread even by their illustrators: Vanessa Bell reportedly said of Woolf's *Three Guineas,* for which she was designing the cover, "I've not read a word of the book—I only had the vaguest description of it and what she wants me to do from Virginia."[59] Most readers who purchased novels doubtless valued the narratives. My point is not that novels' contents ceased to matter but that the novel's meaningful components now also included the visual qualities of its layout and its other material features (paper type, binding, and so on).

"SAY I'M STILL REAL": REPRESENTING THE SUBJECT AS NARRATIVE AND INFORMATION IN VIRGINIA WOOLF'S *ORLANDO*

Novelistic emphasis on textual materiality opposed the medial ideology of information. This had ramifications not only for discussions about media such as microform but also for conceptualizations of the subject as a collection of data rather than an embodied entity. Woolf's *Orlando* exemplifies how a novel might intervene in these discussions. The degree to which a novel could capture its subject (and therefore compete with representations of the subject as data) mattered to the genre's archival project broadly, but it also held an intensely personal significance for Woolf. *Orlando* presents a veiled depiction of her lover, Vita Sackville-West, who had started a series of affairs with other women. Writing *Orlando* provided Woolf with a chance to recuperate the loss of Vita's exclusive affections, "stag[ing] the loss of Vita while simultaneously producing her . . . as biographical object."[60] *Orlando* considers multiple ways in which an individual might be represented: as biographical data, as literary description, as photographs and other images, and as the material remnants and traces produced by life lived in a body. In short, Woolf explores how the print novel can become a multimodal information system, comingling all of these modes, but she ultimately emphasizes the book's material form in order to insist on the irreducibility of embodiment.

Orlando is a singular work in several respects. The eponymous character lives for centuries and changes gender, the literary style is an uneasy alliance of biographical and novelistic conventions, and the tone ranges from glib to earnest. While much about the novel is in flux, one constant is Woolf's exploration of representation and its limits. Although *Orlando*'s tone ranges broadly and is often arch and satirical (Woolf initially described it as "half laughing, half serious"), the author also seems "serious in her attempt to capture the reality of Vita Sackville-West."[61] Elizabeth Cooley observes that when Woolf later composed a biography of her friend Roger Fry, her frustrations echoed "very seriously the concerns that the biographer [narrator] voices quite flippantly in *Orlando*."[62] She asserts command over her relationship with Sackville-West by writing her as Orlando, alternately mocking and revering her, controlling her but also granting her literary immortality.[63] When Woolf, on finishing the novel, wrote to Vita, "What are you really like? Do you exist? Have I made you up?," Vita responded, "I *won't* be fictitious. I won't be loved solely in an astral body, or Virginia's world. So write quickly and say I'm still real." (The letter is signed "Your adoring and perfectly solid Orlando.")[64]

The question of what kind of representation would make Vita "real" tapped into wider discourse about the growing regimes of quantification and personal data collection. Much of this apparatus had formed over the course of the previous century. Mary Poovey argues that the rise of "technologies of representation" in England in the nineteenth century enabled the creation of a new version of national identity: "The image of a single culture had begun to seem plausible in 1860 . . . because the technologies capable of materializing an aggregate known as the 'population' had been institutionalized for several decades. These technologies included the census, which was first conducted in England in 1801, and statistics, which had begun to be institutionalized in the early 1830s. These technologies of representation represented [previously disparate groups] as belonging to the same, increasingly undifferentiated whole."[65] By the twentieth century, the processing of large demographic data sets was increasingly automated: punch cards were used to process census results starting in 1890 in the United States and 1911 in the United Kingdom.

These practices underpinned a conceptualization of the individual as an amassment of data, as "the [early-nineteenth-century] techniques of file-based self-administration . . . constitute[d] the subject as an individual and a citizen."[66] By the time Woolf began writing *Orlando,* people living in the United Kingdom and the United States generated informational records about their lives to a historically unprecedented extent. Buckminster Fuller's Dymaxion Chronofile experiment, begun in 1917, demonstrates how the accrual of information could become a representational act. Fuller's goal was to construct a complete informational archive of a single life—a "faithful comprehensive record" of existence.[67] He would use himself as the "guinea pig," "documenting the life of an individual born in the 'Gay Nineties' . . and maturing during humanity's epochal graduation from the nineteenth century . . . into the twentieth century." He saw himself as the product of an epistemic societal shift, but his ability to create the Chronofile was also indebted to the epistemic shift in information culture. He accumulated a massive archive of paper materials, including "over 37,000 articles written and published by others about [him] or [his] work." By 1980, his Chronofile had expanded to "737 volumes, each containing 300–400 pages."[68]

Fuller's Chronofile illustrates several points about personal information in the early twentieth century. First, while some of the documents he collected, such as newspaper clippings, reflected his status as a public figure, others were the kinds of records that any of his contemporaries might have produced: letters, bills, traffic tickets, and so on. Second, the project's ambitious scope was made manifest by the sheer size of its paper archive. Today, the Dymaxion Chronofile materials housed in the R. Buckminster Fuller Papers at Stanford University take up more than 370 linear feet of shelf space.[69] Such dizzying scale (and physical unwieldiness) is precisely why some of his contemporaries argued that informational documents should be stored as microform. Fuller was acutely aware of just how much information could be generated by a single life because he had the paper trail to prove it. Third, for Fuller, documents and information allowed for self-representation and self-reflection: "I sought to 'see' myself as others might and to integrate that other self with my self-seen self and therefore to deal as objectively as possible with the comprehensively integrated self."[70] In other words, for him, the philosophical aspiration of knowing oneself could best be realized through the accumulation of informational records.

It is within this historical context that Woolf undertook her fictionalized biography of Vita Sackville-West. *Orlando* might be read as a rejoinder to Fuller's assertion that one could gain self-knowledge by amassing informational records. Woolf explores the relationship between biography—a genre wherein facts are narrativized and thus moved into the domain of data—and the novel.[71] She was not the only modernist writer to consider how the novel competed with other forms of life writing and life documentation. Stein lamented that "people now know the details of important people's daily life unlike they did in the Nineteenth Century. Then the novel supplied imagination where now you can have it in publicity." As Stephens argues, Stein held that the novel had become passé (she declared, "The novel as a form has not been successful in the Twentieth Century") because of the proliferation of "the ambient information of the media" and "the instantaneous global transmission of information."[72] Woolf's attitude toward the novel was more optimistic, but *Orlando,* too, bears the traces of the information culture within which the author composed it.

Although *Orlando* is typically classified as a novel, the work is subtitled "A Biography" and features a biographer as its narrator. By throwing the accuracy and integrity of biography into doubt, Woolf exposes the limits of a life represented through information. One problem with such representation is that seminal events may lack documentation, either because they are not recorded or because they contain a full scope of experience not registered by informational accounts. When Orlando falls into a mysterious trance, prefiguring the trance that will precede his transformation into a woman, the narrator states that "the biographer is now faced with a difficulty which it is better perhaps to confess than to gloss over. Up to this point in telling the story of Orlando's life, documents, both private and historical, have made it possible to fulfill the first duty of a biographer, which is to plod . . . in the indelible footprints of truth. . . . But now we come to an episode [that] . . . is dark, mysterious, and undocumented; so that there is no explaining it."[73] Orlando's sex change is the novel's biggest mystery, and the biographer is at a loss to account for it or for the other apparently inexplicable events in Orlando's life.

Although this moment in *Orlando* occurs centuries before the advent of modern data collection, Woolf makes it clear that the fundamental issue is the inability of documentary accounts to record the essence of their subjects. In her essay "The New Biography," she writes, "If we think of truth as something of granite-like solidity and of personality as something of rainbow-like intangibility and reflect that the aim of biography is to weld these two into one seamless whole, we shall admit that the problem is a stiff one." "The New Biography" suggests that biographers are actually less able to represent life than novelists are: they may be "stimulated to use the novelist's art . . . to expound the private life," but they ultimately have "neither the freedom of fiction nor the substance of fact."[74] The biographer-narrator of *Orlando* agrees: "To give a truthful account of London society . . . is beyond the powers of the biographer or the historian. Only those who have little need of the truth, and no respect for it—the poets and the novelists—can be trusted to do it, for this is one of the cases where the truth does not exist" (141). Woolf argues that biography is severely limited by its dependence on fact. The data generated by census records, personal files, and other documents are similarly unable to represent the more phantasmic elements of life. For example, the 1911 census record for

Virginia Woolf only somewhat resembles the entity she became in literary history: "Adeline Virginia Stephen," "Single" (she would marry Leonard Woolf the following year), age twenty-nine, with the listed occupation of "journalist."[75]

Still, Woolf does not uncritically assert the novel's triumph where biography and information fail. In *Orlando,* literature is similarly unable to grasp these elusive essences. Within the narrative, one of Orlando's ongoing quests is to craft a poem that faithfully represents the oak tree on her estate, a project Woolf parallels with the narrator-biographer's task of describing Orlando's life. Orlando believes that literature grants immortality: while "his ancestors . . . and their deeds were dust and ashes," an author could live on because the "man and his words were immortal" (60). Orlando views literature as an information medium, preserving the author along with the subjects represented in the text, in stark contrast to the ephemerality of the physical body. Yet each of the representative quests in *Orlando* is beset by problems, and these problems undermine literature's ability to preserve its subject.

Orlando, for instance, communicates with her husband Shelmerdine in a secret code sent via telegraphs, "a cypher language which they had invented between them so that a whole spiritual state of the utmost complexity might be conveyed in a word or two without the telegraph clerk being any the wiser." To one telegram, Orlando "add[s] the words 'Rattigan Glumphoboo,' which . . . described a very complicated spiritual state" (208).[76] The scene parodies the secret code that Vita and her husband used in letters.[77] It also alludes to the common practice of encoding telegraphic correspondence for privacy. Compare a telegram sent by James Joyce: "MACILENZA PAVENTAVA MEHLSUPPE MOGOSTOKOS." Kevin Birmingham glosses the telegram's meaning: "Private and reliable information received. Number in dispute has been issued. Pending receipt of your letter matter will remain over."[78] Orlando's application of codes to spiritual matters seems perverse, as if "very complicated spiritual state[s]" could be reduced to boilerplate language as easily as business dealings could. As Patrick Collier writes, the scene "dramatizes the benefits and costs of modernism, whose experimental language may represent modes of experience that evade conventional language, but at the cost of a loss of accessibility, of communicability."[79] Shortly after stating how "precisely" their code could describe reality, Orlando is frustrated to realize the difficulties of representing life in literary language: "Life? Literature? One to be made into the other? But how monstrously difficult!" (209). Literature is information in *Orlando,* an encoding of reality into textual form. But life is hard to quantify, and literature, like language, eventually highlights the gaps between the representation and its subject: "Nature and letters seem to have a natural antipathy; bring them together and they tear each other to pieces" (14). *Haptic Storage*

Woolf hoped that a writer who could blend biography with novelistic narrative might have more success. While composing *Orlando,* she praised the biographical work of Harold Nicolson (Vita Sackville-West's husband) as indicating a new direction that life writing might take: "Mr. Nicolson has proved that one can use many of the devices of fiction in dealing with real life. He has shown that a little fiction mixed with fact can be made to transmit personality very effectively."[80] *Orlando* similarly blends the fictional with the real. Woolf had visited Knole (the basis for Orlando's estate), and she drew from *Knole and the Sackvilles,* Vita's nonfictional account of her estate and ancestors. She wrote *Orlando* during a period in which she explored life writing from many angles, producing and publishing two essays on biography, two fictionalized biographies (*Orlando* and *Flush: A Biography*), and a more traditional biography (*Roger Fry: A Biography*) in the space of thirteen years.[81] *Orlando*'s biographer is a comic figure; consider his reductive description of Orlando's change of sex as a "simple fact" (103). Still, he also borrows Woolf's language, stating that "Nature . . . [makes] us . . . unequally of clay and diamonds, of *rainbow and granite*" (59, emphasis added). While *Orlando* undermines the separation between documentary data and literary representation, the novel's insistence on the limitations of both confirms Woolf's ultimate diagnosis in "The New Biography": Nicolson, whose work she admired, could not "mix the truth of real life and the truth of fiction. . . . Although both truths are genuine, they are antagonistic. . . . Let it be facts, one feels, or let it be fiction; the imagination will not serve under two masters simultaneously" (233–34). Neither biography nor the novel bridges the gap between subject and representation. Here the storage capacity of the book hardly matters. If language fails to encode its subject—if textual representation fails, regardless of its scale or genre—an entire library of microform documents would be inadequate.[82]

BODY, HEADER, SPINE: EMBODIED LIVES, TACTILE PAGES, AND THE BOOK'S METAPHYSICS OF PRESENCE

In *Orlando,* neither literary descriptions nor the "factual" documentation of biography can unproblematically grasp the nuances of a life, particularly the mysteries of that life as it is lived in a human body. Yet Woolf proposes that a novel might be able to capture something of this embodied life by mobilizing the embodiment of the print book. *Orlando* demonstrates that a novel can become a material archive, preserving the physical traces left when a reader interacts with a book. These traces may be indexical impressions, like fingerprints, or small objects left between the pages. I refer to such preservation as *haptic storage,* after Laura Marks's "haptic criticism." As she explains, "haptic criticism is mimetic: it presses up to the object and takes its shape."[83] Similarly, haptic storage is created when a subject literally presses up against the medium, leaving a mark. These traces are important because they are nonrepresentational. They conflate embodied subjects with the traces they leave rather than represent them as narrative or data. Haptic storage is possible in the print book because the reader touches the surface of the textual medium directly instead of reading the text through a mediating interface.

The haptic register of books was salient for Woolf, given her work as a printer. Virginia and Leonard Woolf founded the Hogarth Press in 1917 and operated it together until 1938. Leonard's initial motivation for purchasing a press was the physical nature of printing, which "was . . . to be a form of therapy for Virginia," "a manual occupation . . . [that] would take her mind completely off her work."[84] Virginia took on the typesetting, work she found so consuming that her handwriting began to suffer.[85] Nancy Cunard recalled the Woolfs warning her about printing's messy tactility: "I can still hear their cry: 'Your hands will always be covered with ink!'"[86] Woolf's experiences as a printer influenced her novelistic use of the page. Laura Marcus and Julia Briggs have argued that the use of blank spaces in *Jacob's Room* was influenced by Woolf's typesetting of experimental modernist poetry.[87] John Lurz contends that she systematically directs the reader's attention to the page and the book in *Jacob's Room* and *The Waves,* a strategy he interprets as linking the experience of reading to mortality and the passage of time.[88] While the textual layout of *Orlando* is fairly conventional, Woolf disrupts its unmarked status in several places, most notably with the insertion of "a great blank" space, a visual gap mimetically representing the notion that "the most poetic [conversation] is precisely that which cannot be written down" (186). While the move is Shandyean (Woolf credits Sterne in *Orlando*'s acknowledgments), it also mimics the unusual spacing of modernist poetry. Most importantly, it shows *Orlando* to be metamedial, aware of its textual artifactuality. As a consequence, it draws a connection between the diegetic manuscripts and books within the narrative of *Orlando* as they accumulate the traces of their readers and *Orlando*'s own function as an object that might acquire such marks through the course of its reading.

Haptic storage features in several of Woolf's writings. Jacob's eponymous room contains "a Greek dictionary with the petals of poppies pressed in silk between the pages," while in *The Waves* Bernard "press[es] flowers between the pages of Shakespeare's sonnets."[89] Thomas Lewis suggests that, "for Woolf, objects (a book, a house, a

chair) and individuals carry the facts of their history with them. . . .
Even a book accumulates something of the sensibility of each person
who reads it."[90] This is the critical turn of haptic storage: the wear a
book incurs by being read can "summon . . . ghosts," as Woolf wrote
in "The Countess of Pembroke's Arcadia," turning a novel into an
object that preserves through touch. It is because the reader sees and
handles the surface of inscription directly—because she reads the
text without a mediating interface—that the print novel is capable
of haptic storage in a way that a microform Readie could not be. The
form of the medium that gives shape to textual information allows
the accrual of another, material, layer of information.

Orlando features two main examples of haptic storage. The first
occurs via the manuscript of Orlando's poem, "The Oak Tree" (based
on Sackville-West's prize-winning poem The Land), which Orlando
spends centuries rewriting. Orlando is dissatisfied with her attempts
to represent the tree in poetry. But the manuscript of the poem is
itself a record. Its damage chronicles adventures from Orlando's life:
"she had carried this [manuscript] about with her for so many years
now, and in such hazardous circumstances, that many of the pages
were stained, some were torn" (172). These stains—the manuscript
is "sea-stained . . . , blood-stained, travel-stained"—document her
journeys, a mute but potent alternative to the narrator's biography
(172). The manuscript is nearly sacred to Orlando, who reaches for "her
bosom . . . where the pages of her poem were hidden safe. It might
have been a talisman she kept there" (121). The language is echoed later:
"Then Orlando felt in the bosom of her shirt as if for some locket or
relic of lost affection, and drew out no such thing, but a roll of paper"
(172). Like a relic, the manuscript has deep affective resonance. Aura
animates it: the manuscript is "shuffling and beating as if it were a
living thing" in her bosom (200). The text it contains is secondary to
its signification as a physical object.

Orlando's second example of haptic storage again invests a text's
physical wear with preservative value and aura. Orlando goes to her

chapel to read from a prayer book, "a little book bound in velvet, stitched in gold" (127). In an echo of "The Countess of Pembroke's Arcadia," when Orlando holds this book, she feels linked to its previous readers. These links are established through the debris these readers have left in the book. First is "a brownish stain," detectable to "the eye of faith" and said to be the blood of Mary Queen of Scots, who had "held" the book; the word conflates ownership with touch (127). Orlando examines the items in the book and contributes her own: "In the Queen's prayer book, along with the blood-stain, was also a lock of hair and a crumb of pastry; Orlando now added to these keepsakes a flake of tobacco, and so, reading and smoking, was moved by the human jumble of them all—the hair, the pastry, the blood-stain, the tobacco—to such a mood of contemplation as gave her a reverent air suitable in the circumstances" (127–28). There is an element of bathos in Orlando's being moved by mundane items, yet Woolf also emphasizes the text-object's sacredness. Orlando sits in a chapel, opens a prayer book, and is filled "with all the religious ardor in the world" (128).

This scene contrasts the book as container of linguistic information with the book as container of physical objects, attributing its value to its capacity for haptic storage. Orlando does not read the prayer book; she meditates on the remnants it holds. Elizabeth Freeman interprets Orlando's interaction with the prayer book as "erotohistoriography," a practice that uses "the body as a tool to effect, figure, or perform [the] encounter" "between a lost object and a present moment." In Freeman's reading, Orlando's interaction with the prayer book is sexualized: "Lovingly fondling Queen Mary's prayer book, Orlando adds a flake of tobacco to the hair, bloodstain, and crumb of pastry already stuck to its pages. These marks of use are all connected to the body, which seems to variously shed, bleed, eat, and smoke a sedimented history that interests Orlando far more than the textual materials enclosing it." These erotics, which pursue "a visceral encounter between past and present figured as a tactile meeting," are inextricable from the larger questions of informational representation and preservation that inform the medial sensibility with which Woolf stages these haptic moments.[91] The bloodstain in the prayer book links this document with Orlando's manuscript, which is similarly stained (and similarly auratic). The book becomes the vehicle through which social exchanges of readers (including lovers) can occur.

Woolf's descriptions of the prayer book weave real book history into her fictional account of haptic storage. Queen Mary reportedly gave the prayer book to Vita's ancestor Thomas Sackville, on whom Woolf based some details of Orlando's life. Madeline Moore reports that the book "was still at Knole when *Orlando* was written."[92] While there are conflicting reports on the book's whereabouts, Woolf's characterization of the prayer book as "a little book bound in velvet, stitched in gold" is visually accurate (127).[93] If the book was at Knole, Woolf herself might have held it. In this object, then, Woolf combines the history of a real book, a literary representation of that book, and a largely fictional account of the traces left in that book. I say "largely" because I suspect that Woolf was partly inspired by a visit to Knole, where she imagined Vita's dead "ancestresses" materializing behind her: "All the centuries seemed lit up. . . . [A] crowd of people stood behind, not dead at all; . . . & so we reach the days of Elizabeth quite easily. After tea, . . . [Vita] tumbled out a love letter of Ld Dorset's (17th century) with a lock of his soft gold tinted hair which I held in my hand a moment. One had a sense of links fished up into the light which are usually submerged."[94] In this real episode, Woolf linked a lock of hair kept in a letter (another example of haptic storage via a tactile textual medium) to her feeling that the ghosts of Vita's relatives were present. Compare the lock of hair in Orlando's prayer book. *Orlando* locates the power to preserve and evoke the past in the book, a medium whose materiality seemed meaningfully tied to human bodies and human presence.

As Woolf describes it, haptic storage retains a degree of embodied presence that eludes representation and quantification—especially in the case of books. Orlando's poetry cannot capture the living oak; the biographer cannot capture Orlando's complex existence. This latter point is most apparent when Orlando becomes a woman. The biographer frantically alternates between reconstructing facts from official documents and writing an elaborate literary masque. Both attempts are farcical and reductive: Orlando's change of sex is the mystery of the novel, and neither can make sense of it. It is precisely Orlando's embodied life that language cannot hold. If the body evades language, however, Woolf reminds us that a book can store the blood of a human body. Haptic storage draws on the long-standing trope of metaphorizing the human body as a book and vice versa, paralleling embodied lives and embodied texts. Amaranth Borsuk writes that much of "the language we use for the codex suggests its corpus."[95] Allison Muri notes that "the analogy of human bodies as textual, written documents was a very old one, and the image of the human body's creation as text was widespread." Muri outlines the many ways in which "our pages and our bodies have long converged in metaphor": "A page has a body, a header, and a footer; it might contain an appendix or index (from the Latin meaning 'indicator' or more specifically 'fore-finger') or footnotes or frontispiece (from the late Latin *frontispicium,* from *fronts,* 'forehead' and *spic-,* denoting 'see'); it might be part of a chapter (from the Latin *caput,* head); it may be part of a manuscript (from *manus,* hand) or it may be bound into a book with a spine."[96] Early books were made from the skins of animals, and the rectangular shape and vertical orientation of the page may have been determined by human anatomy, designed to fit the hand.[97] While writing was for centuries theorized in terms of absence, this metaphoric tradition instead constructs book and page as signifying the body's presence.

Without wanting to reify this bookish metaphysics of presence, which is no less illusory than the perception that microform makes information immaterial, I note its influence in the history of the book. Because books and bodies have been intimately connected in book design, book production, and reading, words printed on a page bound in a book may provide a more apt, or at least a more immediately apprehensible, vehicle for metaphorizing the body. This likeness, along with the book's affordance of preserving the body's material traces in the manner I have been describing, reinforces the belief that a book can store a subject's embodied presence more readily than other storage media can. Woolf invokes the book-body metaphor explicitly: when Queen Elizabeth I meets Orlando, "she read[s] him like a page" (19). We need not take metaphor literally; my point is not that a drop of blood or a lock of hair actually does preserve the essence of a living human being, nor that such objects are necessarily better than the information we receive from narrative description or documents. It is that because the print novel can accommodate all of these modes, it can offer unique insights into theorizations of literary representation as well as assumptions about the relationship among information, media, and form that often go unexamined.

One point remains to clarify: how the more auratic aspects of haptic storage may inflect *printed* texts. Because neither Orlando's manuscript nor the illuminated prayer book is an artifact of print culture, their auratic quality might appear to stem from their being handwritten. *Orlando* repeatedly critiques print books, associating them with a mercenary publishing industry and depicting them as flimsy and impersonal. Consider Orlando's first visit to a modern bookstore: "All her life long Orlando had known manuscripts; had held in her hands the rough brown sheets on which Spenser had written in his little crabbed hand. . . . But these innumerable little volumes, bright, identical, ephemeral, for they seemed bound in cardboard and printed on tissue paper, surprised her infinitely" (208). The uniformity of these mass-produced volumes contrasts them with the singularity of manuscripts. Woolf's characterization of print here aligns with Walter Benjamin's in "The Work of Art in the Age of Mechanical Reproduction." For him, the problem with "the technique of reproduction"—which includes printing—is that "by making many reproductions it substitutes a plurality of copies for a unique existence."[98] This plurality negates the possibility of authenticity, because "the presence of the original is the prerequisite to the concept of authenticity," as well as of aura, which accumulates based on a work's "presence in time and space, its unique existence at the place where it happens to be."[99] By this logic, the printed volumes that Orlando encounters are incapable of acquiring the aura that she associates with her manuscript and prayer book. Moreover, because these printed books are cheaply made, they seem unlikely to persist to store either information or physical traces.

Haptic storage dismantles this opposition between the hand-written and the printed. Of course, neither category is monolithic. Handwriting might be used in literary manuscripts, private correspondence, business documents, or other contexts. The category of print was similarly expansive, incorporating cheap, mass-produced books as well as fine printing, methods such as lithography and steel-die and copperplate printing, and newer technologies such as photo-offset and near-print. Any of these techniques could produce textual documents capable of haptic storage, provided that the process of reading them involved manipulating the medium directly. That print is not wholly different from handwritten texts is indicated by the similar language Woolf uses to describe the prayer book and the printed edition of "The Oak Tree" that Orlando finally publishes. Orlando's publication is "a little square book bound in red cloth"; the prayer book, "a little book bound in velvet," was in reality also red (237, 127). This language may also be a metamedial reference to *Orlando:* Woolf wrote that "Orlando will be a little book."[100]

Even cheaply printed books become individuated when they preserve traces of their readers. Benjamin argues that "the changes which [a work of art] may have suffered in physical condition over the years as well as the various changes in its ownership" contribute to its aura.[101] A personal copy of a printed book might acquire such a history. Orlando initially feels wary about the uniform, printed books she encounters, but she buys "everything of any importance in the shop" (209). Benjamin, too, collected books. For him, collection was about owning books (not reading them), yet he is in agreement with Woolf that they acquire meaning in their relationship to an individual: "Not only books but also copies of books have their fates. And in this sense, the most important fate of a copy is its encounter with [the collector], with his own collection."[102] Susan Stewart has also noted this affinity between books and their owners: "The royal predilection for giving libraries the names of their benefactors . . . has in more modern times between transferred to the identification of the reader with the books he or she possesses, to the notion of the self as the sum of its reading."[103] For Benjamin, the collector transfers part of his soul into his objects: "Ownership is the most intimate relationship that one can have to objects. Not that they come alive in him; it is he who lives in them."[104] There are differences between Benjamin and Woolf, but they agree on the potential for a print book to become significant by preserving some sense of its owner. Woolf's intervention is to link the perception that a book can mysteriously preserve its reader's aura to a materially grounded consideration of the book's preservation of physical traces.

Like Orlando with her prayer book, the reader of *Orlando* personalizes her copy as she reads. Intentional personalization was quite common in the nineteenth century, when, as H. J. Jackson has detailed, readers would frequently insert "miscellaneous bits and pieces" into books, creating extra-illustrated copies.[105] *Orlando*'s prayer-book scene was ironically echoed in the extra-illustrated copy of *Orlando* owned by Sackville-West's mother: "Vita's mother read it and was horrified; she thought it a vicious and hateful book. . . . She put a photograph of Virginia in her copy of the book and wrote next to it: 'The awful face of a mad woman whose successful mad desire is to separate people who care for each other. I loathe this woman. . . .' She covered her own copy with outraged graffiti."[106] Vita's mother used *Orlando* to rewrite Woolf as Woolf had rewritten Vita, allowing her to create a record of her antagonism. But the ability to touch, mark, and manipulate books also makes it possible for books to generate networks of readers, configured across time by touching the same object.[107] Collier writes that "one way [Woolf] critiqued the literary public sphere was by constructing imagined exchanges between readers and writers that were more private, more intimate" than those in the literary market.[108] In *Orlando,* this takes the form of poetry functioning as "a secret transaction, a voice answering a voice" (238). A hand might also answer a hand through the transaction of reading.

Orlando helped to construct and contextualize the relationship between Virginia Woolf and Vita Sackville-West, not only through its narrative but also as a material object. Karyn Z. Sproles writes that "*Orlando* occupies the midpoint of the Woolf / Sackville-West literary exchange, occurring within the multiple context of a relationship that ran persistently on three tracks: personal contact, private correspondence, and literary publications."[109] The Hogarth Press printed thirteen of Vita's books. Her *Sissinghurst* is a physical record of the pair's complicated relationship, bearing the text "By V. SACKVILLE-WEST" and "Printed by hand by Leonard & Virginia Woolf" on the title page and the dedication "*To V.W.*" on the first page.[110] *Orlando* has a corresponding dedication "To V. Sackville-West." Vita felt an emotional connection to Virginia through books. She wrote to Woolf: "Today as I was driving . . . I saw a woman . . . carrying [the novel To] the Lighthouse. . . . I saw your name staring at me, Virginia Woolf. . . . I got an intense dizzying vision of you: you in your basement, writing; you in your shed at Rodmell, writing; writing those words which that woman was carrying home to read."[111] Regarding the woman almost jealously, she makes clear the book-object's power to bear the presence of another. Virginia and Vita had often sent each other books as gifts. Vita's first gift to Woolf was a copy of *Knole and the Sackvilles,* and Virginia presented Vita "with the leather bound single volume manuscript of Orlando."[112] *Orlando* itself became the relic of the beloved, testifying to the way in which books have been understood to store and communicate presence.

Microform reading machines and hand-printed books defined the opposite ends of a spectrum of ideas about the embodiment of information and media in the first decades of the twentieth century. *Orlando* explores how print novels may represent their subjects in several ways, combining literary description and biographical information with the physical traces accrued by their tactile medium. Metamedial novels usefully contest the split between the informational and the material that so frequently characterized writing about information media during the modernist period. To speak of information storage is typically to mean the preservation of content rather than objects. Archivists classify items as having informational value or artifactual value. For the former, the primary importance lies in the information they contain. For the latter, it lies in their physical existence—their aura, perhaps. An object in an archive may have both informational and artifactual value, and each value register has implications for how different media mediate information.

Both microform's medial ideology of disembodied information and the book's metaphysics of presence are illusory. This should not stop us from confronting how such beliefs are constructed and activated or how they influenced debates about information's management and mediation in the last century. The idea that books are connected to the presence of living people (whether authors or readers) is long-standing, due to a confluence of factors ranging from haptic storage to Christian traditions of the incarnate word.[113] Every information medium has a history of design and use that encourages certain assumptions and attitudes. Metamedial novels such as *Orlando* can clarify the extent to which the transparency or hypermediacy of information media influence these perceptions. I do not take Woolf to be valorizing the novel or the book in *Orlando* so much as reflecting on the limits of representation and on the part that media play in theorizations of representation. No medium offers perfectly comprehensive or lossless preservation. A flake of tobacco cannot grant immortality any more than a biography, a poem, or an archive of documents can. Woolf's intervention is to highlight the richness of archival undertakings that synthesize multiple modes of storage.

When media transitions spark claims of the print book's obsolescence, not all concerns about how reading or accessing information will change are the product of Luddism or uncritical nostalgia. Books are sensuous, visual, tactile—even olfactory. Cunard recalled how "the smell of printer's ink pleased [her] greatly, as did the beautiful freshness of the glistening pigment. . . . After a rinse in petrol and a good scrub with soap and hot water, [her] fingers again became perfectly presentable; the right thumb, however, began to acquire a slight ingrain of gray, due to the leaden composition."[114] Printing left its traces on Cunard's hands, as the hands of future readers would leave traces on the books she printed. Even microform's staunchest advocates acknowledged the book's value. Microform proponents may have envisioned a future in which information was liberated from the constraints of the book, but the book continued to dominate their imagination and shape their rhetoric. Throughout his career, Otlet returned to the book as a metaphor for new information systems. In addition to conceptualizing indexed documents as components of a "single universal book" of knowledge, he referred to his ideal total information system as the "Biblion."[115] He would continually refer to new technologies as "substitutes for the book," a practice already evident in his 1906 paper hailing microform as "a new form of the book." While the book remained useful as a conceptual framework for Otlet, the book as a physical object held value for Fiske. His memoir documents his bibliophilia: "I have always felt more at home in a library than in any other place. Men of my name have been identified with books for many generations, and my earliest recollections are of lying on the floor in my father's library reading books."[116] Brown—who published popular fiction and traded rare books—also had a more complex relationship with books than his Readies rhetoric indicates. A quotation from *The Readies* is telling: "With reading-words freely conveyed [by machine] maybe books will become as rare as horses after the advent of the auto, perhaps they will be maintained only for personal pleasure or traditional show" (40). In other words, if books were inadequate for the scale of modern information management, they still possessed qualities worth pursuing in and of themselves.

Microform and its reading machines, avant-garde typography, cheaply produced bestsellers, and fine-printed novels constituted a literary media ecology. Taken together, they inspired a variety of ideas about how the embodiment of media, and the ways in which that embodied form was mediated, affected the ability to store information. For all its apparent precarity, the book continued to dominate debates about how people should manage information, whether dealing with information on the scale of the entire world or of the individual life. Even Brown wanted to leave his own mark—his "fly-speck"—in books. As he writes in *1450–1950,*

> I like to see
> fly specks
> on yellowed pages
> I like too
> leaving my own on
> new ones.[117]

Although the title of this collection of poems points to eras that Brown imagined as being beyond the dominance of print (he writes of "the illuminated mss. [manuscripts] Of / 1450" and "the / more illuminating / moving scripts of / 1950"), he reflects here on the physical traces that books acquire.[118] His poem portrays the book as a medium that preserves traces accumulated over its history—traces that each reader can leave, interacting directly with the printed surface—and reminds us that leaving even the most trivial of marks can be pleasurable and meaningful.

Bodies of Information

Digital Immortality and the Corporeality
of Books

Books were once such handsome things. Suddenly they seem clunky, heavy,
almost fleshy in their gross materiality. Their pages grow brittle. Their ink
fades. Their spines collapse. They are so pitiful, they might as well be human.
> —Ben Ehrenreich, "The Death of the Book"

By recording your life digitally you have the opportunity to bequeath your
own ideas, deeds, and personality to posterity in a way never before possible.
With such a body of information it will be possible to generate a virtual you
even after you are dead.
> —Gordon Bell and Jim Gemmell, Total Recall: How the E-Memory
> Revolution Will Change Everything

GHOSTS IN THE MACHINE

In 1993, Penguin published English writer Peter James's novel *Host*
in an unusual format: a pair of floppy discs. Slightly more than two
decades later, London's Science Museum put *Host* on display, bill-
ing it as the first electronic novel.[1] Although largely forgotten today,
Host created a media sensation, and James was "accused of killing the
novel."[2] A quarter-century after its initial publication, it appears to
be at once outdated and prescient. E-reading has evolved from discs
inserted into desktop computers—a platform so unsuited to pleasure
reading that one journalist performatively brought "his computer on a

wheelbarrow to the beach, along with a generator" to read *Host*—to portable, dedicated e-readers.[3] It has flourished since the launch of the Amazon Kindle in 2007, and the current digital literary landscape fulfills the prediction James made when he published *Host*: "I said at the time that electronic novels would catch on when they became as nice to read as the printed page and more convenient than a printed book. That day is more or less here, and I'm pleased to see that far from killing the novel, it is thriving more than ever."[4]

I begin this chapter on information mediation and the twenty-first-century print novel with a novel from the 1990s because *Host* illustrates an issue that more recent writers and thinkers, from novelists to computer scientists, have continued to grapple with: how the transition from books to digital media, with their distinct methods of mediating information, have affected beliefs about death, mourning, and the preservation of life. *Host,* which sensationally transformed the novel from a print book into a digital file, is a science-fiction thriller about the quest for digital immortality. In it, Professor Joe Messenger helps Juliet Spring, a terminally ill woman with whom he has become infatuated, escape death by downloading her consciousness into a computer system. Already displaying signs of mental instability before this transformation, Spring becomes unhinged as a distributed, networked consciousness: she cyberstalks Messenger's family, demanding he find her a new body. This digital novel, in other words, explores the consequences of digital embodiment. James portrays human minds and digital information as structurally interchangeable, but *Host* concludes that there is something monstrous about digital immortality.

The computational mediation and management of digital information has had important ramifications for theorizations of the relationship between identity and embodiment. I examine how the forms that digital information takes as it is processed by computers and accessed through screens have strengthened the belief that information exists independent of its medium and that, consequently, an individual's mind might be converted into digital information. While books have long been metaphorically associated with the corporeal aspects of being, digital information and digital media have underpinned a posthumanist conceptualization of the subject as a disembodied intelligence. This has resulted in the imagination of digital afterlives as well as pragmatic attempts to realize them through digital technology. I analyze several digital resurrection projects that assert that people can transcend the mortality of the human body by being transformed into digital information, whether by preserving a digital representation created from personal information accumulated about that subject or by downloading that mind into a computer. I show that these projects implicitly or explicitly reject paper and books as storage media whose limited scale and physical presence are at odds with digital transcendence.

Throughout this book, I have been arguing that, in the modernist era and again in the twenty-first century, writers have produced metamedial novels that emphasize the formal structures of the book in order to insist on the ways in which information's organization and embodiment (which are obscured by modern regimes of mediation) are nontrivial and constitutive of meaning. When it comes to digital immortality, the study of the metamedial novel opens up productive avenues of inquiry about how books and other media have been implicated in ideas about embodiment's relationship to both information and being. One response by contemporary novelists to the possibility of digital immortality has been to reinvest in the connection between subjectivity and embodied existence through an emphasis on the corporeality of books. Digital immortality threatens to usurp literature's traditional role as a vehicle of immortality. Shakespeare's Sonnet 18 famously articulates the hope that literary representation could guarantee eternal existence: "So long as men can breathe, or eyes can see, / So long lives this [poem], and this gives life to thee." As we saw with *Orlando,* this hope has animated the novel's archival project. Writers have used the novel's scale, informational qualities, and print materiality as strategies for representing and thus perpetuating their subjects. But posthuman subjects might persist instead as digitally mediated information, literal ghosts in a machine. As such, digital immortality would also make another novelistic impulse obsolete: to work through grief and loss via the intertwined forms of narrative and the book.

My aim in this chapter is to demonstrate the degree to which perceptions about information media's embodiment have contributed to rhetoric and ideologies concerning the nature of the self as well as to ideas about how that self might be represented through information, novelistic narrative, material traces embedded in media, or a combination of these modes. I analyze two metamedial novels that illustrate the tensions between these modes of representation. Steven Hall's *The Raw Shark Texts* (2007) and Jonathan Safran Foer's *Extremely Loud and Incredibly Close* (2005) consider how the materiality of the book remains relevant to the work of mourning at a time when human consciousness (or its digital approximation) might persist beyond existence in the body. Alongside these novels, I analyze Anne Carson's poetic elegy *Nox* (2010), which similarly turns the book into a monument through which grief can be processed. As Hall's and Foer's fictional protagonists, like Carson, struggle to make sense of the bereavement each has suffered, their attempts to recover traces of their lost loved ones negotiate the archival capacities of narrative and of the print book in a culture that increasingly values abstracted information and digital storage media (and in which books are also being transformed into digital entities). The difference between Carson's undertaking and that of Hall and Foer lies in the novel's potential to cope with loss via fiction, using narrative as a way of coming to terms with death and trauma. Together, these literary works illustrate how the print book's tactile materiality, in contradistinction to the mediations of digital information media, has underscored a model of identity rooted in embodied presence. This chapter documents the range of ways in which a metaphysics of presence has operated (or been negated) in the digital media ecology, has been emblematized by the book's textual materiality and connected to how information media foreground or obscure their own forms, and has been brought to bear on attempts to bridge the gap between the living and the dead.

DATA AVATARS, DIGITAL INFORMATION MEDIATION, AND THE POSTHUMAN SUBJECT

Until recently, the notion that digital technology could provide a means of transcending death was confined to the realm of fiction. Digital immortality has been a staple plot element in science fiction, from cyberpunk novels to television to film. Fictional narratives of digital immortality tend toward one of two tropes. In the trope of downloaded consciousness, living subjects become immortal when their brain is transferred into a computer. A consciousness might then be downloaded into a robot (as in the *Star Trek: The Next Generation* episode "Schizoid Man") or another human body (as in *Altered Carbon*) or, most commonly, might persist within a computer or a network (as in William Gibson's *Neuromancer* and James's *Host*). In the trope of digital resurrection, the original individual dies but leaves behind a digital re-creation amassed from a synthesis of personal information. Others can then interact with this data avatar. (Narratives featuring this trope include the *Black Mirror* episode "Be Right Back" and General Electric's fiction podcast "LifeAfter.") Both tropes equate an individual with the representation of that individual as a collection of digital information, which can exist beyond the original context of the human body.

Fictional narratives of digital immortality predate the age of Big Data, but they have flourished in an epoch defined by the scale, ubiquity, and automation of data collection and analysis—as has the work of researchers who hope to achieve some form of real digital immortality by taking advantage of these affordances of digital media. The shift from print media to digital computers as the dominant medium for information management made the imagination of digital immortality possible even before these undertakings became feasible. This occurred first because digital media enable the collection and analysis of quantities of personal data on an unprecedented scale. These new thresholds make realistic text-based data avatars viable; they are also of such magnitude that they appear to approach the horizon of complete and total informational representation. Second, the ways in which digital information is mediated by digital media and computational systems has given rise to the perception that information (particularly in digital forms) is essentially immaterial, lending support to the claim that consciousness is both commutable into information and transferable from a body to a hard drive. Finally, the transition from paper and books to computers as primary information storage media has instated a new set of media metaphors for understanding the individual subject: whereas books' apprehensible materiality inspired a set of metaphorical connections with human bodies, computers have been likened to the workings of the human mind. This change in metaphoric register has resulted in a deemphasis on embodiment as a necessary component of identity.

Digital resurrection technology has expanded rapidly, drawing on research in natural language processing, machine learning, artificial intelligence, and other fields. Interaction with the simulated digital consciousness is usually limited to conversation via a text or voice interface rather than through an artificial body.[5] These data avatars are essentially chatbots. For example, the startup company Luka develops chatbots, including "an AI [artificial intelligence] friend" named Replika designed to comfort people who are "feeling overwhelmed."[6] (The appeal of this potentially uncanny technology is evident in the fact that, within three months of Replika's 2017 launch, more than 2 million users had signed up.)[7] Replika's application of chatbot technology originated with the unexpected death of Roman Mazurenko, a close friend of Luka co-founder Eugenia Kuyda. Grieving her loss of him, Kuyda transferred Mazurenko's text messages to her (and to many of his friends and family, with their consent) into a neural network, creating a chatbot as a "digital monument" to her friend.[8] (At the time of this writing, the "Roman Mazurenko" app is still available from the iTunes store.) Another example is Augmented Eternity, developed by entrepreneur and academic Hossein Rahnama. The technology uses a person's emails, social media posts, and other digital textual communications to build an AI avatar. Although "your physical being may die," Rahnama says, "your digital being will continue to evolve with the purpose of helping people and maintaining your legacy as an evolving being."[9]

Such projects are possible because of the quantities of information people generate using digital communication technology and because of software that facilitates the management and aggregation of this information. Consider the sea change in personal data collection that has taken place since Buckminster Fuller began his Dymaxion Chronofile. Although the scale of Fuller's archive was impressive for his era, his practices of writing and archiving have been superseded (and dwarfed in scale) by the automatic harvesting of personal data that is an inescapable feature of online communication. As Frank Pasquale writes, "new hardware and new software promise to make 'quantified selves' of all of us, whether we like it or not. The resulting information—a vast amount of data that until recently went unrecorded—is fed into databases and assembled into profiles of unprecedented depth and specificity."[10] Governments and corporations engage in "dataveillance"; many individuals also voluntarily track their own data—for instance, through fitness trackers and gamified habit-tracking apps. As Victoria Vesna puts it, "in the computerized paradigm, humans are perceived as information . . . or as information-processing entities."[11]

This scale has become a prerequisite not just for imagining what a digital afterlife might look like but for building an operable version of it. For Rahnama, Augmented Eternity is possible because of the amount of personal data produced today: "You couldn't do this in the past [because] you didn't have enough of a digital footprint about someone to be able to do these types of sophisticated predictions. . . . Fifty or 60 years from now, millennials will have reached an age where they have collected zettabytes . . . of data individually."[12] A zettabyte is equal to 1 trillion gigabytes. As a point of comparison, if each of Buckminster Fuller's painstakingly amassed 260,000 pages of personal archive were equivalent to one page in a Microsoft Word document, Fuller would have accumulated roughly four gigabytes worth of data.[13] As Rita Raley argues, the scale of digital data is significant not only because of its vastness but also because of its interconnectedness across platforms and institutions: as "large-scale data-aggregating corporations . . . and increasingly sophisticated tracking technologies such as Flash cookies and beacons indicate a shift in scale," "the linking of databases, corporate actors and institutions . . . radically changes the scope of a query."[14]

This change in scope and interconnectedness, which moves digital data, according to Raley, into the realm of the "incalculable," is predicated on the dominance of digital media.[15] For Kuyda and Rahnama, the information that animates an avatar is all digital—text messages, social media posts, electronic mail. The most emphatic example of the centrality of digital media to digital immortality research is MyLifeBits. Microsoft researcher Gordon Bell began this life-logging project in 1998, and it was active through 2007.[16] He pursued his goal of creating "a lifetime store of everything" by wearing cameras to record his interactions, digitizing his paper records, and taking digital photographs of himself and his possessions.[17] In a decade, Bell collected more than 350 gigabytes of personal data—relatively modest compared to the zettabytes Rahnama predicts, but nearly a hundred times what Fuller was able to collect over a lifetime.[18] While MyLifeBits had a number of aims, its most radical was to use a person's extensive personal archive of information to create an interactive avatar: "With such a body of information it will be possible to generate a virtual you even after you are dead. Your digital memories, along with the patterns of fossilized personality they contain, may be invested into an avatar. . . . Your digital self will reach out to touch lives in the future, allowing you to make an impact for generations to come."[19] Perhaps a Freudian slip, the phrase "a *body* of information" reveals the gap between the self living in a body and the digital simulation of that self.

The substitution of digital texts and images for paper records and personal artifacts was crucial to Bell's vision. "Set a goal of being paperless within a year," he declared; "besides scanning the paper you already have, you should also arrange to receive more born-digital communications in the future." Digital data take up less space to store and are more easily managed by computing contexts that can make use of them, from algorithms to neural networks. For Bell, however, the desire to go paperless went beyond pragmatism. He described "paperless offices" as "far more pleasant, and somehow more calming," and he scanned and destroyed his paper records with zealousness: "nothing beats the feeling of feeding your paper to the shredder and seeing your clutter evaporate."[20] Embedded in the ethos of MyLifeBits is a disregard, bordering at times on antipathy, for paper as a manifestation of the material world's physicality.

The relationship between digital mediation and the perception of information as immaterial and disembodied is not strictly deterministic, however, because this ideology of information predates digital culture. In chapter 3, I showed that the rhetoric surrounding microform's promotion expressed a similar view of information. N. Katherine Hayles has documented how, with the rise of cybernetics in the middle of the twentieth century, "information lost its body," becoming "conceptualized as an entity separate from the material forms in which it is thought to be embedded."[21] In her summation, "the posthuman view privileges information pattern over material instantiation, so that embodiment in a biological substrate is seen as an accident of history rather than an inevitability of life."[22] As she has argued, this conceptualization of information rests on a dualistic, hierarchical dichotomy, splitting mind from body and information from medium and disregarding the many ways in which the body shapes the formation of the self. Ted Striphas has described a similar trajectory whereby information's original formulation as an embodied entity was replaced by the idea that information is abstractable:

whereas early definitions of information centered on spiritual and legal contexts, "both of which locate information in the body vis-à-vis its incarnations, godly or performative," the modern "object-oriented definition" instead "inaugurates a process of abstracting information from the body; instead of being vested there, information becomes a separate raw material."[23]

While this dualistic separation of information and embodiment predates digital culture, the widespread perception of this split as a fundamental feature of information has much to do with the phenomenological aspects of the computational paradigms of the late twentieth and early twenty-first centuries. As Matthew Kirschenbaum has argued, the belief that digital information is less embodied than information stored in media such as books has been pervasive among new media theorists as well as among lay users (although Kirschenbaum's work has done much to correct this misconception in more recent media theory). He describes (and contests) what he calls the medial ideology of digital information: the "commonplace notions of new media writing" including the ideas "that electronic texts are ephemeral" and "that electronic texts are somehow inherently unstable and always open to modification."[24] The net effect is that electronic writing is seen as symbolic rather than inscriptive. These views have been produced by an overvaluation of the screen as a site of computational interaction. Such screen essentialism tends to gloss over the materiality of the hard drive, the procedures of code, and other aspects of computing.[25]

Beyond this issue of a problematic focus on the surface of computing systems, the operations of the underlying algorithms may be so complex that even computer scientists struggle to explain them. The algorithmic processes that power machine learning applications such as Replika and Augmented Eternity can produce sophisticated interactions with digital avatars without the need for either users or researchers to have a detailed mental model of the system's information. As Ed Finn writes of the virtual assistant Siri, the ability of artificial intelligences "to interpret real-world commands depends on two key factors: natural language processing (NLP) and semantic interpretation. . . . The major breakthroughs in algorithmic speech analysis have come by abandoning deep linguistic structure—efforts to thoroughly map grammar and semantics—in favor of treating speech as a statistical, probabilistic challenge."[26] The issue is once again that modern information mediation entails the use of interfaces (whether screens or procedures) that remove the need for users to understand or interact with the physical forms information takes as it is stored, the organizational systems that map semantic relationships within a mass of data, or the processes that make use of data.

One final respect in which the shift from paper and books to digital information media has intersected with rhetorical constructions of the posthuman subject as disembodied information comes through the metaphors that print and digital media have respectively inspired. As I described in chapter 3, the corporeality of print books, combined with the history of book design and books' capacity for haptic storage, granted them a conceptual kinship with human bodies. The result was a metaphoric tradition of describing human bodies via the bodies of books, and vice versa. Computers expanded the metaphors through which people are conceptualized via media: digital systems are frequently discussed in terms of human cognition, and vice versa. The idea that consciousness and the brain are akin to—or, in fact, are forms of—digital information or computation has been a consistent assertion, from popular writing about neuroscience to research into complex systems and artificial intelligence. Finn traces the development of the theorization of human cognition as computable as part of an intellectual history of abstraction and universal principles, reaching back to Gottfried Leibniz: "Many complex systems demonstrate computational features or appear to be computable. If complex systems are themselves computational Turing Machines, they are therefore equivalent: weather systems, human cognition, and most provocatively the universe itself."[27] If cognition is a form of computation, this argument goes, it must be an information-processing system; and information, especially information in digital form, is easily transferred. The abstraction of computation dovetails with the abstraction of information.

Compare the philosopher Nick Bostrom's description of the processes by which some researchers hope to attain downloaded consciousness, which stresses the interchangeability of the structures of a human brain and computational structures: "First, create a sufficiently detailed scan of a particular human brain. . . . Second, . . . reconstruct the neuronal network that the brain implemented, and combine this with computational models of the different types of neurons. Third, emulate the whole computational structure on a powerful super-computer. If successful, the procedure would result in the original mind, with memory and personality intact, being transferred to the computer."[28] If "thought is a much broader cognitive function depending for its specificities on the embodied form enacting it," as Hayles writes, consciousness cannot be extracted from the brain without being altered.[29] Yet the assertion that consciousness is formally similar to digital information and computable systems has been widespread. Digital information media and computational systems offer a means of potentially realizing some version of digital immortality, even as their mediations and management processes create an understanding of information as fundamentally disembodied, reinforcing the notion that one can be reduced to a collection of digital data in the first place.

MOURNING AND THE METAMEDIAL NOVEL: MATERIAL BOOKS, MIMESIS, AND STEVEN HALL'S *THE RAW SHARK TEXTS*

Novelists who want to engage with contemporary conversations about death, immortality, and representation through their metamedial novels must contend with a range of issues surrounding information and media. As the novel's role as an archive is challenged by digital immortality research that promises to represent its subjects with an unprecedented degree of fidelity—digital avatars as literal mimesis—the print book's material form puts it at odds with the medial conditions that have contributed to downloaded consciousness and digital resurrection. The book's affordances highlight the centrality of information's embodiment to its meaning, modeling information as inextricable from its form. Moreover, the print book stands in not only for embodiment and tactility broadly but for the human body specifically—that container that posthumanists seek to discard. As such, the metamedial print novel would seem to be precluded by posthumanism's turn toward digital information.

An additional issue is that books, like posthuman subjects, are also becoming digital. In the 1990s, the spread of the World Wide Web made the electronic editions of Project Gutenberg accessible to readers; in that same decade, the first generation of dedicated e-readers (including the SoftBook and Rocket E-Book Reader) launched, albeit with little success. E-reading became a commercially viable alternative to print in the twenty-first century with the launch of the Sony Reader (2006), the Amazon Kindle (2007), and the Barnes & Noble Nook (2009). E-book sales rose dramatically: between 2008 and 2010, net unit sales for e-books increased by more than 1,000 percent in the United States, while sales for paperbacks fell by nearly 14 percent.[30] In 2010, Amazon announced that its Kindle editions were outselling hardcover books.[31] Between 2010 and 2014, the percentage of Americans who owned an e-reader spiked from 4 percent to 32 percent.[32] Tablet computers also increased the market share of e-books, as did the availability of the Kindle app on mobile phones. These trends were not stable during the next half-decade—e-book sales fell sharply in the United States and the United Kingdom in 2015 and 2016 and continued declining in 2017—but the viability of e-reading was established.[33] Novels have been especially affected by the so-called e-book revolution. Forty-four percent of all American e-book sales in 2017 were in adult fiction, larger than any other category; in the United Kingdom, fiction accounted for 50 percent of e-book sales in 2016.[34] Of the twenty top-selling e-books published in the United States in 2017, eighteen were novels.[35]

As Amaranth Borsuk writes, "digital reading devices existed prior to e-readers, but none has challenged our definition of the book in quite the same way or raised as much of an outpouring of print nostalgia on the one hand and digital futurist rhetoric on the other."[36] One response from the publishing industry has been a renewed emphasis on book design, with the reasoning that readers are more likely to purchase print editions over e-books if the former add aesthetic value. In his acceptance speech for the 2011 Man Booker Prize, Julian Barnes remarked on his novel's design: "Those of you who've seen my book—whatever you may think of its contents—will probably agree that it is a beautiful object. And if the physical book, as we've come to call it, is to resist the challenge of the e-book, it has to look like something worth buying and worth keeping."[37] Barnes's slippage from novel to book suggests the close links between the two, while his statement reinforces the idea that the book's material presence is one of its defining features. In contrast, e-books mediate information in ways that downplay the presence of informational texts: Borsuk notes that while e-readers "remediate those [features] that developed with the print codex," "the[ir] design . . . has gradually streamlined to minimize buttons and dials, heightening the sense that they are simply interfaces for engaging with text and perpetuating the myth of digital disembodiment."[38] (That this myth connects e-books with posthumanist visions of the self as disembodied information is evident in Bell's recommendation that people interested in pursuing his model of digital immortality "invest in Amazon's Kindle, Sony's eBooks, or another similar reading device.")[39] Theorizations of the print book's fetishization or relegation to the realm of nostalgia in the digital era, such as Kiene Brillenburg Wurth's "book presence" and Jessica Pressman's "aesthetic of bookishness," turn on a similar emphasis on the book's materiality—on "the book as artifact."[40]

I am interested in examining how the book's hypermediate materiality is in tension with the apparent immateriality of digital information—and how both influence theorizations of the relationship between mind and body in the production and representation of identity. Metamedial novels illuminate this tension, one that has informed how a number of contemporary writers have leveraged the book's form to posit alternatives to posthumanist constructions of the subject as divorceable from the body. Many contemporary metamedial novels, including Reif Larsen's *The Selected Works of T. S. Spivet* (2009), Salvador Plascencia's *The People of Paper* (2005), and J. J. Abrams's and Doug Dorst's *S.* (2013), have explored how the presence of the book can be used to mourn or memorialize. Writing of the book's persistence, John T. Hamilton suggests that "perhaps it is simply the book's material presence that we are hesitant to abandon: . . . a sense of the real in contrast to the merely virtual."[41] Metamedial novels can explore how fictional narrative and the book's material form may provide a means of working through loss or continuing to feel connected to those who have died.

Steven Hall's *The Raw Shark Texts* exemplifies this strategy: the novel portrays digital immortality as monstrous, proposing that both narrative mimesis and the book's ability to retain and re-create the presence of the subject are better ways of connecting with the dead because they are rooted in attempts to preserve or reproduce physical embodiment. In Hall's novel, the protagonist, Eric Sanderson, tries to recuperate two losses. He suffers from extreme dissociative amnesia, a loss brought on by the accidental death of his partner, Clio Aames. His quest to recover his memories (and identity) and his lost love is tied to the question of which medium is best able to preserve them. Eric encounters a variety of information media within the narrative, from typed letters and print books to computers and databases. As a literary artifact, too, *The Raw Shark Texts* foregrounds media and mediation: the novel has more than forty pages of visual material, including photographs, a film still, and a scanned newspaper clipping, as well as a series of typographic images of the eponymous shark. These images culminate in a flipbook section in which the shark appears to swim toward the reader. The novel is also part of a media network comprised of elements that include an alternative reality game and the distribution of "missing" episodes from the narrative.

While *The Raw Shark Texts* acknowledges that print is dependent on digital media for aspects of its production, the novel systematically problematizes digital storage technologies, instead asserting the unique relevance of the storage properties of text inscribed on paper—particularly for coping with loss and death. Hall depicts paper and books as having a greater degree of embodied materiality than digital information does, in that the former can be held, touched, and manipulated at the site of inscription. He focuses his examination of the book's materiality on haptic storage, documenting how paper media capture impressions and indexical traces. Because they are nonrepresentational, such traces stand as an alternative to the representation of the subject as a collection of information. It is not enough for Eric to discover information about Clio; he wants to resurrect her as a living, breathing woman. *The Raw Shark Texts* suggests that print media are better suited to this task because of their more apprehensible material presence and the way in which they preserve the physical traces of their readers.

The question of how differently various information media preserve their subjects features prominently in the novel. On the first page, Eric awakens with no idea of who or where he is. He contacts the psychologist Dr. Helen Randle, who explains that his memory loss is the result of a fugue state triggered by the trauma of Clio's death. While Eric has regained his memory after previous amnesiac episodes, the fugue state has recurred nearly a dozen times, with increasing severity. But a far stranger explanation for the amnesia soon arises. Eric receives letters signed by "The First Eric Sanderson," an incarnation of himself predating the current fugue state.[42] The letters explain that his memory loss is caused by a cryptozoological creature called the Ludovician. The Ludovician is a "conceptual shark," an animal that lives in streams of information and feeds on human memories. It is destructive *and* preservative, a vicious creature thought in myth to store anyone whose memories it has consumed. The Ludovician is also emphatically textual, manifested in Hall's typographic imagery.

As Eric travels in search of the shark, aided by a conceptual fish expert named Dr. Trey Fidorous and a young woman named Scout, the narrative consistently figures digital media as dangerous. A letter from The First Eric Sanderson cautions, "The internet, remember there is *no* safe procedure for electronic information. Avoid it at all costs" (81). Hall himself seems to take this advice, as *The Raw Shark Texts* gives digital media only passing mention and relatively generalized description. One of the novel's few direct references to computer code comes with a description of text that "appears to be programming source code," but this code is not presented as operable or machine-readable (96). It comprises broken bits of text, printed and arranged to resemble a mosquito (much as typography is used to construct

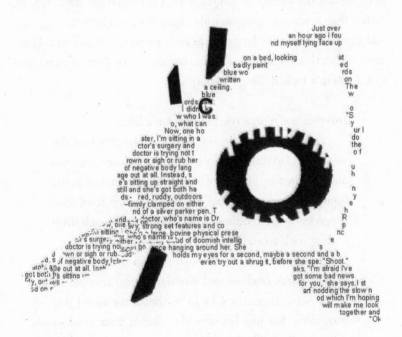

FIGURE 4. The Ludovician. Image by Steven Hall. Reprinted with permission.

images of the Ludovician). Far more attention and detail are given to pre-computing forms of code, including Morse code and a substitution cipher that uses a typewriter keyboard. This emphasis on predigital media is also evident in the book's visual and kinesthetic effects. The typographic images recall experiments in concrete poetry, and the flipbook recalls the kineograph, first patented in 1868. While Hall's use of visuals is one of the most notable attributes of *The Raw Shark Texts,* demonstrating the book's ability to incorporate other media, the postcard message, the newspaper clipping, and the typographic imagery all reinstate written and printed text on paper as the novel's preferred communication media.

Although paper's materiality is crucial to *The Raw Shark Texts's* argument about how media preserve their subjects, the novel also considers how narrative representation may work toward that end. It expresses the dream of language in a prelapsarian state, where, rather than acknowledging an unbridgeable chasm between signifier and signified, narrative language perfectly encodes its subject. Hall's most direct treatment of this theme comes with a sheet of paper that Eric finds in a locked filing cabinet:

> Imagine you're in a rowing boat on a lake. . . .
>
> You reach over the side and feel the shock of the water. . . . Holding out your hand, you close your eyes and feel the tiny physics of gravity and resistance as the liquid finds routes across your skin, builds itself into droplets of the required weight, then falls, each drop ending with an audible tap.
>
> Now, right on that tap—stop. Stop imagining. . . . Here's what's obvious and wonderful and terrible all at the same time: the lake in my head, the lake I was imagining, has just become the lake in your head. It doesn't matter if you never know me. . . . I could be dead, I could have been dead a hundred years before

> you were born and still . . . the lake in my head has
> become the lake in your head. (54–55)

This treatise on literary representation argues for a direct correlation between the subject and its representation (and therefore its transmission and preservation): "the lake in my head has become the lake in your head." Eric finds this passage printed on paper; the implication is that such idealized representation might be possible with print— which is, of course, the primary medium of the novel.

In this fantasy, printed text is capable of not only diegesis but also mimesis: the word on paper embodies the signified instead of merely describing it. As Laura Marks puts it, "mimesis is a form of representation based on getting close enough to the other thing to become it."[43] This connection between printed narrative and mimesis is strengthened when Eric is given a glass "full of thin paper strips," each of which has "the word WATER printed on it in black ink" (285). In order for Eric to enter the conceptual world of the shark, he must first "drink the *concept* of the water" (285). When he is finally able to do this, the printed text actually becomes the represented subject: "The paper and the words were gone. Now there was . . . real physical water where the words had been" (307). Hall likens this process to the doctrine of transubstantiation. When Eric drinks the concept of water, he thinks, "*The wine becomes the blood of Christ*" (286). The novel portrays print narrative as uniquely able to "represent" water in two senses—to describe it (as in the lake document), and to "re-presence" it, restoring its original material form. Whereas digital immortality proponents envision a future in which people's consciousnesses live on by escaping their bodies, Hall imagines that print narrative might be capable of such complete mimesis that it could preserve or restore the dead while retaining their connection to their bodies. In this latter vision, mimesis occurs both through narrative representation and through the medium's physical presence being exchanged with that of the subject.

The novel acknowledges that this perfect, doubled mimesis is only a fantasy. Narratives are repeatedly shown to be inaccurate and uncertain. Language's general tendency toward approximation rather than accuracy is signaled by Hall's frequent use of words such as "like" (occurring in the sense of "similar to" more than four hundred times) and "thing"/ "something"/ "anything" (occurring more than eight hundred times). A further demonstration of narrative's inaccuracy is evident in "The Light Bulb Fragment," a journal account written by the original Eric, describing the vacation in Greece during which Clio drowned. The account finally grants Eric (and the reader) access to some of his lost memories, but it also expresses concern about its accuracy: "There's no way to really preserve a person when they've gone and that's because whatever you write down it's not the truth, it's just a story" (413). The novel's ending also closes on a highly ambiguous note, casting doubt on the preceding narrative. The last page of the narrative proper describes Eric and Scout (who may be Clio's reincarnation) united in the shark's conceptual world, en route to a Greek island. This is followed by a final chapter, titled "Goodbye Mr Tegmark," that contains only a newspaper clipping ("Body of Missing Man Found") and a postcard of a still from *Casablanca,* with a note addressed to Dr. Randle from Eric ("Whatever happens, please don't feel bad. I'm well and I'm happy, but I'm never coming back" [427]). The reader is left wondering whether the entire story has been the fantasy of a delusional man who could not cope with the trauma of his lover's death. Yet if Eric has died (it is his body that the clipping describes), how does the reader explain the postcard to Randle, which bears Greek stamps? Hall's reference to the astrophysicist Max Tegmark hints that quantum physics might resolve this ambiguity, but the two textual accounts that end the novel—the printed newspaper and the note typed on the postcard—are fundamentally conflicting.[44] Like the Rorschach inkblots echoed by the novel's title, *The Raw Shark Texts* supports conflicting interpretations.

As the novel makes clear, narratives are imperfect records, yet *The Raw Shark Texts* also critiques apparently objective informational accounts on the grounds that abstracted information fails to preserve a subject's physical presence. Eric engages in a "violent hunt for [his] own reference material," attempting to assemble his life from fragments of personal information, hoping to build it into a totality that could either restore his identity or allow him to access some remnant of Clio, who is doubly displaced, dead and forgotten (20). Yet as Eric rues his "life as a shopping list"—a haphazard collection of facts—he realizes that it is precisely embodiment that is lost when a human is condensed into information (30). None of the information he recovers about Clio allows him to reclaim her *"weight in the world"* (42). Hall's most vivid example of the shortcomings of the posthuman subject-as-information comes with Mycroft Ward, the novel's chief villain. Ward achieves the posthumanist goal of transcending his human body: he distills his personality into a data set and uses hypnosis to imprint these data onto a host. Ward eventually operates a network of hundreds of hosts, connected and synchronized through a computer database. Given that his name echoes "Microsoft Word," Ward's story stages a critique of corporate imperialism as well as digital word processing.[45] But even as Ward strives for total information storage, his story highlights information loss. His monstrosity emerges as he migrates to the digital form of a "huge online database" (204). It is when he moves beyond a single human body—a move that he coordinates on a large scale using digital technology—that he becomes a posthuman monster, a "thing" (204).

Bodies of Information

Ultimately, *The Raw Shark Texts* portrays text on paper as better able than digital media to preserve a subject—and thus connect with the dead—not because of their respective representational capacities or their scale but because paper's propensity for haptic storage allows it to capture indexical traces and impressions. Scars, footprints, fingerprints: as surface markings created by the pressure of a body or object, impressions register movement and touch. When Eric expresses his desire to know "a real Clio Aames," he primarily measures this reality in terms of her body ("her *actual skin, voice,* ideas, *eyes,* . . . *blood, fingernails* and shoes and *periods* and *tears* and nightmares, *teeth* and *spit* and *laugh*") and the indexical traces her body leaves ("her *actual fingerprints*") rather than information about her (31, emphasis added). In one of the novel's key scenes, he is finally able to connect to and therefore mourn Clio when he encounters her handwriting and other traces in a guidebook: "an entire galaxy of biro stars, pen orbits and ink loop rings. . . . My fingers touching the indentations, the pen marks, the folded page corners. . . . Clio Aames's real and true and actual writing right there in front of me. . . . Then I noticed my face, my cheeks were wet. . . . I hadn't even realised I was crying" (266). This language describing Clio's handwriting as "real" and "actual" recalls Eric's earlier attempts to grasp "the idea of a real Clio Aames" with "actual skin" and "actual fingerprints." Instead of her body, he has come into contact with the impressions her body has left in a book.[46] As if to emphasize that this scene of his encounter with Clio's indexical traces is the closest possible thing to perfect mimesis, the sound of Eric's tears on the page ("Tap. Tap. Tap.") echoes the tapping of water described in the lake document (266). As Hayles writes of *The Raw Shark Texts,* "in contrast to the dematerialization that occurs when data are separated from their original instantiations, entered as database records, and re-instantiated as electronic bits, inscriptions on paper maintain a visible, even *gritty* materiality capable of interacting directly with human bodies."[47] The book's affordance of haptic storage is newly meaningful in an information culture that prizes information over embodiment and digital media over print and paper.

The Raw Shark Texts distinguishes texts on paper from their digital counterparts because of the way in which the former encourage haptic storage, which, in turn, contributes to the metaphysics of presence associated with paper and books. Reading Hall's novel, Wurth argues that "in literary criticism binary oppositions between electronic and paper-based writing are not very helpful. . . . There is an inevitable, and indeed productive, trace of the one within the other."[48] *The Raw Shark Texts* is a complex media artifact, and it demonstrates that print and digital technologies are interconnected—from Hall's use of Microsoft Word to compose the typographical images to the scene in which Eric writes in the air with a magic calligraphy brush that, when touched to a computer printer, produces printed pages. The novel's embrace of the book's materiality goes beyond the medial ideology: it is not simply that texts on paper are material and electronic information is not (although the novel often imagines this to be the case); rather, the tactile nature of this materiality is as meaningful as the informational content is.

While print books tend to foreground their mediality and materiality, digital information media tend to feature interfaces designed to meet ideals of invisibility and transparency. With digital media, operating systems and screen interfaces intervene between users and the stored information so that the forms information takes as it is inscribed on the device are inaccessible. Users are further distanced by the hermetic sealing that makes information stored on a hard drive so durable. As Jacob Rabinow puts it, random-access disk storage is like a "book [that] can be read without being opened."[49] Because the site of digital inscription is inaccessible, it cannot act as a substrate capable of registering fingerprints or other traces. Contrast Rabinow's readable but untouchable "book" with Eric's musings on preservation: "I thought about how a moment in history could be pressed flat and preserved like a flower is pressed flat and preserved between the pages of an encyclopedia. Memory pressed flat into text" (36). When a flower is pressed in a book, it is preserved not by representing it as a description or a collection of data but by using the book's weight to flatten it. Hall's narrative of memory, mourning, and loss argues that, in an age of digital information, the print book still matters as a storage medium because it can be the means of other kinds of record making. Challenging the posthumanist disregard of embodiment, *The*

Raw Shark Texts explores how the print novel can attempt to reclaim embodied presence, creating a model of literary immortality through its combination of narrative representation and haptic storage.

PAPER MONUMENTS: PRESENT BOOKS AND ABSENT BODIES IN JONATHAN SAFRAN FOER'S *EXTREMELY LOUD AND INCREDIBLY CLOSE* AND ANNE CARSON'S *NOX*

Jonathan Safran Foer's *Extremely Loud and Incredibly Close* is another metamedial reflection on the value of fictional narrative and the material presence of books and paper for those dealing with grief. The novel tells the story of Oskar Schell, a precocious nine-year-old who is trying to cope with the loss of his father, who died in the World Trade Center on September 11, 2001. When Oskar discovers a key hidden among his father's possessions, he sets out on a quest to discover its purpose, hoping to learn more about his father. His narrative is interwoven with those of his grandparents, who emigrated to the United States after surviving the Dresden bombing in World War II. Whereas *The Raw Shark Texts* emphasizes the power of narrative mimesis to evoke presence, *Extremely Loud and Incredibly Close* explores how narrative's fictiveness can help those struggling with loss and trauma to control, and thus cope with, the narrated events. It also asserts the importance of paper and the book as media: in the novel, these media merge with and substitute for the bodies of the dead. Against the posthumanist ideal of the immaterial self constituted through information, Foer proposes the self encountered through fiction and touched through books.

Oskar is only able to begin grieving when he has confronted the gulf between his absent father and information about his father. Initially, he is determined to recover any information that he can about his father's life and death. Reeling from his traumatic loss, he fixates on facts and knowledge: "Etymology is one of my *raisons d'être*," he explains, "which is a French expression that I know."[50] These statements of his knowledge ("that I know") are scattered throughout the book. Yet he is haunted by the one fact that eludes him: although 9/11 is arguably "the most documented event in human history," Oskar does not know the precise circumstances of his father's death.[51] He craves the certainty of fact: "If I could know how he died, exactly how he died, I wouldn't have to invent him dying inside an elevator that was stuck between floors, . . . and I wouldn't have to imagine him trying to crawl down the outside of a building, . . . or trying to use a tablecloth as a parachute" (257). Oskar compulsively tries to piece together his father's final moments, replaying the phone messages his father left while trapped in the World Trade Center and searching the Internet for information. When he finds an envelope in his father's closet labeled with the word "Black" and containing a key, he makes it his mission to contact every New Yorker with the surname Black until he discovers the key's purpose. The quest becomes Oskar's *raison d'être,* as he and his neighbor A. R. Black (one of the first in the phonebook) travel around New York. Even if he cannot learn the details of his father's death, he can perhaps solve *this* mystery and feel as though he can "stay close to [his father] for a little while longer" (304).

Oskar repeatedly finds informational documents that are reductive, imprecise, or irrelevant. For instance, A. R. Black has a card catalogue in which he keeps records on important people. The cards are both brief and comically idiosyncratic: "Mahatma Gandhi: war"; "Tom Cruise: money"; etc. (157). Oskar is incensed to find that his father's life is not recorded in this barest of archives (but 9/11 hijacker Mohammed Atta's is). Oskar later learns that he has been entered into this catalogue: "Oskar Schell: Son" (286). While his relationship to his father forms a core part of his identity, Oskar's self-description on his business card is far more comprehensive: "INVENTOR, JEWELRY DESIGNER, JEWELRY FABRICATOR, AMATEUR ENTOMOLOGIST, FRANCO-PHILE, VEGAN, ORIGAMIST, PACIFIST, PERCUSSIONIST, AMATEUR ASTRONOMER, COMPUTER CONSULTANT, AMATEUR

ARCHEOLOGIST, COLLECTOR" (99). Newspaper reports about the people who died in the 9/11 attacks are similarly terse: "the lists of the dead in the paper" include items such as "mother of three," "college sophomore," "Yankees fan," along with sixty other entries—including five for "father" (273). Foer suggests that the complexities of people's lives resist encapsulation in official documentation or secondhand accounts.

Nor is more information necessarily the solution. Although dataveillance has increased exponentially since September 11, 2001, individuals (or their grieving children) may not be able to access their digital footprints. Pasquale writes of the massive amounts of data collected by "powerful businesses, financial institutions, and government agencies": "Everything we do online is recorded; the only questions left are to whom the data will be available, and for how long."[52] Additionally, the disappointment Oskar feels at the conclusion of his quest comes from his realization that many aspects of life are unknowable or resistant to datafication. He finally learns that the key belongs to a man who met Oskar's father only in passing; this person recalls little except that Oskar's "father seemed like a good man" (300). Although Oskar fails to uncover much information, his search is cathartic: it allows him to connect with his fellow New Yorkers and finally face his guilt at not answering his father's final phone call. Working through grief, Foer suggests, is a matter of community building and self-reconciliation rather than information gathering.

Foer also considers how photographs and other images compare to textual information in their ability to represent and preserve the world. *Extremely Loud and Incredibly Close* is notable for its use of visual imagery: in addition to overlapping text, a newspaper column annotated in red ink, multicolored handwritten scribbles, and reproductions of objects such as business cards, the novel contains fifty-five pages of photographs. The most striking of these is a flipbook in which Oskar reorders images of a body falling from the World Trade Center so that the man's fall is reversed. The novel's incorporation of these images has been controversial. Dominic Head criticizes Foer's use of imagery, particularly the "falling man" image, as "a gesture that seeks to infuse with affect a familiar form of postmodern jokiness."[53] Aaron Mauro asks, "How will it be possible to understand this image as an aesthetic object alongside the horrifying certainty of this man's death?"[54] The visuality of Foer's novel is appropriate, however, given that *Extremely Loud and Incredibly Close* focuses on the aftermath of 9/11, which, as Birgit Däwes notes, was itself "complicated by the role of the media, and especially visual media": "Televised live and in real time around the globe, the attacks were immediately turned into a media spectacle that decisively shaped the construction of their remembrance. Photographs of the event proliferated and testified to the general need . . . to frame the events in visual narratives."[55] In addition to reproducing the media experience of 9/11, images also offer another way of processing trauma. Sonia Baelo-Allué explains that they "prove especially important in the trauma process since to be traumatized is to be possessed by an image or an event not assimilated or understood at the time."[56]

While images are prominent in the novel, Foer emphasizes their inability to capture the materiality of the objects they portray. This is demonstrated by the many photographs of doorknobs (28, 115, 134, 212, 265). Oskar's grandfather, Thomas Senior, took photographs of his apartment in America, including images of "every doorknob," so that "if anything happened, [the insurance company] would be able to rebuild the apartment again exactly as it was" (175, 174). Yet the photographs can no more ensure the apartment's existence than they can make the objects they represent physically present for the viewer. When Thomas Senior searched for his fiancée during the Dresden bombing, he was severely burned by a hot doorknob (211). The next page displays a picture of a doorknob, as if to verify this story. The reader, however, has already learned that the doorknobs photographed belong to a different, later residence. Nor can readers feel the burning heat of the Dresden bombing through the picture. The juxtaposition of photography and narrative makes it clear that neither restores the original object.

The photographs do point to another method by which one might access the physical existence of an object (or person): by touching paper. The photographs reinforce the artifactuality of the book that Foer's readers hold in their hands. There are two main books referenced within the narrative: Oskar's scrapbook and Thomas Senior's daybook (which the latter uses to communicate, having stopped speaking due to his traumatic past). These books become materially present for the reader as the format of Foer's novel mimics them. For example, Oskar describes flipping through the pages of his scrapbook (51); the next fifteen pages of are photographs (53–67), as if the pages of Foer's novel have become the pages of Oskar's scrapbook. The same occurs with Thomas Senior's daybook. Many page spreads contain only a short phrase written in the middle of each page. The reader simultaneously views these pages as belonging to the daybook (which their content and format match) and Foer's novel (the pages are paginated, and they are printed, not handwritten). What Foer creates is not so much a book that imitates visual or digital media as a book that imitates books, its paper pages standing in for and embodying other paper pages. The novel becomes a metafictional and metamedial facsimile: as Bonnie Mak writes, "the traditional facsimile often shares the same platform as the artefact that it seeks to imitate; namely, a traditional facsimile is a codex that imitates a codex."[57] Foer's meta-book heightens the verisimilitude of the narrative, at the same time reinforcing the reader's sense of paper and the book as storage media whose tangible physicality is central to their meaning.

Whereas Hall considers how paper's accessible materiality can re-create the presence of a person who has touched it, Foer explores how the materiality of paper, taking up space and asserting its felt presence, can supplement or substitute for the bodies of the dead. The material presence of books is evident throughout *Extremely Loud and Incredibly Close*. Books fill a bathtub and a grandfather clock, and they shelter young lovers (179, 127). This physicality can be dangerous when books and paper become fuel for fires. Oskar's grandmother wonders if her collection of letters contributed to the destruction of her house in Dresden: "Sometimes I would think about those hundred letters laid across my bedroom floor. If I hadn't collected them, would our house have burned less brightly?" (83). Oskar thinks similarly about the paper in the Twin Towers: "I read that it was the paper that kept the towers burning. All of those notepads, and Xeroxes, and printed e-mails, and photographs of kids, and books, and dollar bills in wallets, and documents in files [. . .] all of them were fuel. Maybe if we lived in a paperless society . . . Dad would still be alive" (325, ellipses in brackets are Foer's). In the book he published about MyLifeBits, Bell mentions this passage from *Extremely Loud and Incredibly Close,* writing that it "struck a chord" with him as he digitized his paper records. Bell's description of this process demonstrates his characteristic dislike of paper: "I never knew quite how much I'd resented the need to stockpile so much paper until I saw it dwindle away like dirty old winter snow in the spring thaw."[58] But Bell misses the broader point Foer makes about books and paper in the novel: that their physical presence is not simply clutter or kindling, but a means of encountering the presence of the dead.

Oskar seeks his father through the traces of his handwriting on paper. Although some of the handwriting he finds turns out to be from other people, the fact that he *wants* the writing to be his father's shows that tactile interaction with paper can create emotionally significant records. As if to emphasize this close connection between bodies and paper, Thomas Senior seems to be merging with his daybook. His hands are tattooed with the words "yes" and "no"; he bleeds into the pages of his book; and when he cries into his daybook, it is "wet with tears running down the pages, as if the book itself were crying" (120, 180). Oskar and Thomas Senior become able to grieve for Oskar's father when they dig up his empty coffin, fill it with the bundles of letters that Thomas Senior has been writing for decades, and rebury it. Texts on paper, sealed together into a coffin as if bound together in a book, substitute for the missing body at the center of Foer's novel.

No technology is capable of resurrecting Oskar's father or reversing the events of September 11. Yet the forms of paper and the book (rather than digital avatars or massive information stores) finally enable Oskar to process his grief.

Foer's approach to working through loss through the materiality of books and paper is similar to Anne Carson's approach in *Nox*. Merging memoir and elegy with her translation of Catallus's Poem 101, *Nox* is her nonfictional reckoning with the complicated grief she experienced at the sudden death of her estranged brother Michael. *Nox* reproduces the scrapbook Carson made in the wake of his death. Among the remnants this assemblage text incorporates are childhood photographs, torn scraps of a letter, typed passages of Latin definitions, and Carson's commentary on the process of translating the poem (itself an elegy for Catallus's brother). Just as *Extremely Loud and Incredibly Close* turns the pages of Foer's novel into Oskar's scrapbook and Thomas Senior's daybook, giving these fictional codices material form, *Nox* turns Carson's real scrapbook into a simulation. A reader who absently reaches to touch a scrap of tape or crumpled paper will be reminded that *Nox*'s pages only reproduce images of the original book. As Liedeke Plate writes, "the visual, trompe-l'oeil effects . . . are deceiving to the eye but not to the touch."[59] *Nox* evokes the intimacy and presence of Carson's original scrapbook while also emphasizing its absence, much as the entire work grapples with the absence of her brother. If to read *Nox* is to be "drawn into the material world of things," including Carson's personal mementos, "all this 'show' of materiality, authenticity, and autography in *Nox* points . . . to a material presence that is staged, screened, derived, or second-hand."[60] *Nox*'s unusual format also produces this presence-absence dynamic. Rather than being bound into a codex, the pages are attached side by side, folded into an accordion structure that is set within a sturdy box the size of a large book. The pages can be turned from one side of the box to the other, but as they are not affixed to the interior they can be removed from their container. Stretching out to more than eighty feet, *Nox* stresses the book's physical presence, even as this very present corporeal form reminds the reader of Michael's absence.

Carson parallels the impossibility of reproducing the nuances of Catallus's poetry in translation with the impossibility of coming to terms with her brother's absolute alterity. Michael had been distant long before his death. "Travelling on a false passport and living under other people's names"—resisting official attempts to document his existence—he had spoken with Carson "maybe 5 times in 22 years" prior to his death.[61] "There is no possibility I can think my way into his muteness," she writes, echoing her translation of Catallus's elegiac address to his brother as an attempt to "talk . . . with mute ash." The distance she felt from Michael in life is magnified in death, made palpable by the difficulty of memorializing him: "He's dead. Love cannot alter it. Words cannot add to it" (1.0). Death remains an inescapable fact, and it eludes informational records and literary representation alike. As the title suggests (*nox* is Latin for "night"), neither poetic elegy nor memoir can dispel the darkness cast by death. Instead, Carson concludes, the goal is to continually memorialize: the task of translating a poem "never ends" for the translator, and so, too, "a brother never ends" (7.1).

Yet Carson indicates that she is able to achieve some catharsis regarding her brother's death through the process of forming these musings and mementos into a book. In life, Michael was often lost to her, adrift somewhere in the world; after his death, he was cremated and his ashes were scattered at sea. Carson was not present for this event because his widow did not find her contact information until several weeks later. *Nox* functions as a surrogate corpse, a tomb, and a monument to Michael's life. "When my brother died," Carson explains on the back of the box in which *Nox* is packaged, "I made an epitaph for him in the form of a book." Plate describes how, "torn and scattered across the pages of *Nox*, Michael's letter to his mother metonymically re-presents his disseminated ashes," and Wurth argues that "the title page bearing [Michael's] name . . . may be *in lieu* of a tombstone."[62] Emphasizing how a book's pages may stand in for absent corpses, Carson recalls how "both [her] parents were laid out [for their funerals] in their coffins . . . in bright yellow sweaters. They looked like beautiful peaceful egg yolks" (5.5). This description comes two pages after a page displaying a slip of paper on which two ovals have been painted in egg-yolk yellow, as if to symbolize her parents' bodies. "There is no stone," she writes of her brother; but the gray box that houses *Nox*'s pages evokes a tombstone as well as an archival storage box (5.6). The final page features the Catallus poem in its original Latin, blurred beyond legibility as if damaged by water and fixed against a black background—as if the poem has become her brother's ashes, scattered at sea and irretrievable.

As *Nox* stands in for Michael's body and gives his memorialization physical form, Carson prompts the reader to consider how people's interactions with the materiality of paper and books may shape how they conceptualize and make peace with death. Plate argues that "the bookishness of *Nox* involves an ethical command," with "its thingness giving sustained instruction and practice in how to approach otherness," including the otherness of the dead.[63] When a reader expands the accordion pages, *Nox*'s "thingness" is inescapable: as the pages expand out of their container and across a room, they are unwieldy and slightly uncanny, like a corpse lolling out of its coffin. *Nox* is a testament to Carson's ability to turn her grief into an artifact. "It is when you are asking about something that you realize you yourself have survived it," she writes, meditating on the similarity between history and elegy, "and so you must carry it, or fashion it into a thing that carries itself" (1.1). *Nox* becomes that thing, both a monument to grief's endlessness and a means of reconciling with it.[64]

The critical difference between Foer's and Carson's uses of the book as a way of contending with death is that, in *Extremely Loud and Incredibly Close*, the metamedial engagement with the work's medium is tied to the ways in which fiction also aids in the process of dealing with grief. Although *Nox* similarly turns to the book as a way to cope with loss (and similarly focuses on the difficulty of representing a life in language), Carson's work is not a novel. *Extremely Loud and Incredibly Close* weaves together several fictional stories, most prominently Oskar's first-person narration and the letters from his grandparents recalling the traumatic events of their past. Where information on its own is incomplete and unhelpful, Foer shows that narrative organizes that information and gives it meaning, creating a process for working through trauma. Although trauma is often theorized as lying outside conscious knowledge and thus outside language, it is nonetheless linked to narrative: as Cathy Caruth writes, trauma "is always the *story* of a wound that cries out, that addresses us in the attempt to tell of a reality or truth that is not otherwise available."[65] Narrative becomes a therapeutic process whereby the narrator can recontextualize the traumatic events, allowing for their cognitive assimilation.

Although Oskar compulsively invents stories about what might have happened to his father, he tends to seek refuge from difficult emotions in the domain of information. His fixation on the certainty of fact is a coping mechanism, but it is not working well: Oskar's therapist recommends institutionalizing him (206). In contrast, narrative is a coping mechanism that allows for the rehabilitation of traumatic events. *Extremely Loud and Incredibly Close* is one of a number of post-9/11 novels that seek to make sense of the terrorist attacks through fiction.[66] Narratives are not a panacea, of course. Rendered mute by his traumatic past, Thomas Senior writes letters in which he tries to make sense of the death of his pregnant fiancée in Dresden as well as his abandonment of his wife and son—Oskar's father—in America. As Thomas Senior writes, he wishes he had "an infinitely long blank book and the rest of time" (279). The implication is clear: there will never be enough time or room on the page for him to write through all of his trauma. As if to dramatize this, the lines of text on the page become smaller and closer together until they overlap, eventually turning the page into a black, unreadable mass (276–84). This typography, as Hayles notes, imitates handwriting on paper while revealing the digital technology required to create the visual effect; the net result is "a visible indication of the trauma associated with the scene."[67] Narrative is constrained by the limits of media. Moreover, narrative is necessarily incomplete, subject to the same shortcoming as informational records are: both are finite accounts that do not represent the complexity of a lived life.

Although narrative fails to fully capture its subject, it can invent its subject instead. Oskar and his grandmother finally cope with their losses by creating narratives that retell their traumatic events in reverse, as if undoing them—the textual equivalent of Oskar's reordered sequence of images that reverses the trajectory of the man falling from the World Trade Center. His grandmother writes of a dream in which the destruction of Dresden is reversed: "In my dream, all of the collapsed ceilings re-formed above us. The fire went back into the bombs, which rose up and into the bellies of planes whose propellers turned backward, like the second hands of the clocks across Dresden" (306–7). By her story's end, all of history has been reversed, back through the universe's creation (313). Foer's novel concludes with Oskar's fictionalized reverse-narrative of his father's death and of the attacks on the Twin Towers:

> The plane would've flown backward away from him, all the way to Boston. He would've taken the elevator to the street and pressed the button for the top floor. . . . Dad would've gone backward through the turnstile, then swiped his Metrocard backward, then walked home backward. . . . He would've gotten into bed with me. . . . I'd have said 'Nothing' backward. He'd have said 'Yeah, buddy?' backward. I'd have said 'Dad?' backward, which would have sounded the same as 'Dad' forward. . . . We would have been safe. (326)

They were *not* safe, but "Oskar's acceptance of the possibility inherent in his imagination is captured in his conditional modal verb form in the final line of the novel."[68] The narrative reordering of events cannot undo reality, but it grants the narrators some control, allowing them to reconcile themselves to the trauma.[69] Shortly before this scene, Oskar had exhumed and then reburied his father's coffin, filled with Thomas Senior's letters; these two acts, together, provide new closure. Foer indicates that Oskar is beginning to move forward with his life, even as he concludes the novel with Oskar's backward narrative. Narrative conditions how reality is understood and given meaning. In a world where certainty cannot exist—at least, not for some of the events that matter the most—Foer suggests that narrative is at least as fair a reflection of truth as informational documents are, and that narrative allows for the fullest emotional engagement with life.

For all the optimistic futurism of digital immortality research, many of these projects are haunted by the specter of death. The influential futurist Ray Kurzweil discusses his father's death twice in *The Singularity Is Near,* in which he describes the methods he believes will lead to immortality.[70] Bell's friend and colleague Jim Gray, with whom he had devised MyLifeBits, was tragically lost at sea, and Bell's wish "to immortalize [Gray] in the most rich and resilient way possible" drove his subsequent work on the project.[71] Grieving Mazurenko, Kuyda describes a chatbot as being "like a shadow of this person"; to speak to a chatbot is "similar to . . . imagining we're talking to someone we've lost, or even talking to a therapist."[72] A year after Mazurenko's death, she was still engaging in regular conversations with the memorial chatbot.[73] The centrality of grief to so much digital immortality research is striking.

The novelistic response to death I have been describing is to confront and work through that loss. Contemporary writers have emphasized the book's materiality as a way of reinforcing the importance of embodiment to information as well as to human consciousness. Another response is to reimagine and rewrite the events of loss, composing narratives through which one comes to understand that loss or reclaim some connection with the dead. Paper and books are key to both approaches, whether as tactile vehicles for narrative or as media that foreground their material presence. When we outline the ramifications of the book's mediation of information, then, one factor is how the book's hypermediacy prompts readers to contend with embodiment. Once again, the cultural work of the metamedial novel is to explore a range of potential responses to discourse regarding media's relationship to form, information, and representation.

Designers of digital media are increasingly focused on touch as an element in user interfaces. As David Parisi describes,

> Computer scientists, roboticists, engineers, and psychologists . . . have had some significant—and some more modest—successes . . . , formally establishing computer haptics as a new discipline in the 1990s, incorporating vibrating "rumble" motors into more than 500 million videogame controllers . . . , building high-fidelity haptic devices for use in surgical training and remote manipulation, making somewhat effective cybersex machines, developing prosthetic limbs capable of feeding complex tactile sensations back to their wearers, and embedding vibration feedback mechanisms in touchscreen interfaces as a means of approximating the sensations produced by pressing buttons and keys.[74]

In response to Peter Miller's statement that "the digital, far from killing the material world, seems only to have intensified our attachment to it," Willard McCarty asks, "Is that as odd as it might first seem? I question the implicit causality in 'to have intensified'; I wonder whether it would be better, more accurate, to say that the digital has arisen simultaneously with our intensified attachment to the material." After all, he notes, "'digital,' from the fingers (digits) of our hands, evokes touch and manipulation etymologically."[75] As we move toward the Internet of Things, material objects are increasingly imbricated in the information landscape, and digital interfaces are designed with the purported intuitiveness of touch at their forefront.

Yet touch remains a sensory category that resists direct mediation. John Durham Peters characterizes touch as "the most resistant" of all of the senses "to being made into a medium of recording or transmission":

> It remains stubbornly wed to the proximate. . . . Touch defies inscription more than seeing or hearing, or even taste or smell. . . . Though materializing mediums, telephone promoters, and radio performers all tried to transport touch, their efforts at such cloning always fell eerily short. A very different stance toward touch is found in the argument of some poststructuralists that the body is itself a text. As fruitful as this insight can be, it risks missing the skin, hair, pores, blood, teeth, eyes, ears, and bones of these texts, and more important, their short life span.[76]

For these reasons, the loss of touch intensifies the pain of mourning. Virtual interactions may allow comfort—and some grief counselors have proposed that data avatars may help the bereaved in the near future—but, as Peters writes, "being there still matters, even in an age of full-body simulations."[77] Returning to the narrative of digital bodies and digital media with which I began, I note that *Host* concludes with the horror of disembodiment. "I'm not enjoying myself," says Spring's digitized persona. "I feel like a ghost. . . . I can't do anything, can't *feel* anything."[78] Neither the impressions of readers left in books nor the physical presence of the books themselves can compensate for the absence of living, embodied subjects—for their "weight in the world," to quote Eric Sanderson—but they offer another form of contact and another way of making sense of loss.

The print book participates in what Jacques Derrida calls the secret of the archive: the ability of a printed or impressed mark to signify presence through its indexical recording of the moment the inscription device touched the inscribed surface. In *Archive Fever,* Derrida takes Freud's reading of the story of Hanhold in Wilhelm Jensen's *Gradiva* as a parable for how the printed page is impressed. The archaeologist Hanhold becomes enchanted with Gradiva, a woman he sees depicted in bas-relief in an Italian museum. He dreams that he meets her as she walks through the hot falling ashes of Mount Vesuvius, and he is left only with her footprint, preserved in the hardened remnants of Pompeii. For Derrida, the printed page does not so much store the traces of its readers' bodies as the touch that results in its creation: "The singular imprint, like a signature, barely distinguishes itself from the impression. And this . . . is the condition for the uniqueness of the printer-printed, of the impression and the imprint, of the pressure and its trace in the unique *instant* where they are not yet distinguished one from the other. . . . The trace no longer distinguishes itself from its substrate."[79] Gradiva's footprint is "an impression that . . . almost confuses itself with the pressure of a footstep that leaves its still-living mark on a substrate, a surface, a place of origin."[80] Such traces constitute a different kind of record, one that is nonlinguistic and "still-living," conflating the embodied human with the mark she leaves rather than representing that subject as information. If digital immortality is "a potential way to escape the human form," as James puts it in *Host,* metamedial novels invest in an ideal of embodied presence that returns readers to the form of the human body through the form of the book.[81]

Both fiction and haptic interactions with media can temporarily bridge the gap between self and other, living and dead, if only speculatively or indirectly. For contemporary novelists exploring the implications of posthumanism or digital information mediation, the print book is a relevant and multifaceted site for thinking through the intertwining of informational representation and embodiment. A book can become a memento or a monument, or a space where a mourner encounters the traces of a lost loved one.

Shelf Life

Media Transition, the Death of the Novel, and the Futures of the Book

---

The age of print is passing.
> —N. Katherine Hayles, How We Think: Digital Media and
> Contemporary Technogenesis

There is something congenitally troubled about the history of the book. Always at its wit's (if not virtual) end, the book is forever actively engaged in its own disappearing act.
> —Henry Sussman, Around the Book: Systems and Literacy

FLICKERING SIGNIFIERS

A few semesters ago, I was preparing the final lesson for my Introduction to Literary Theory class. My plan was to end the course by showing the students a work of electronic literature and asking them to apply the theoretical approaches they had studied over the course of the term to interpret it. When I opened the electronic poem I planned to teach the next day—Sasha West's "Zoology," with visual design by Ernesto Lavandera—I found that the poem was, for lack of a better term, broken. When functioning, the poem's lines progress through a combination of programmed movement arcs and user interactions. A few phrases from the poem appear and disappear, accompanied by colored dots that break apart and reform in the shape of animals;

after each scene dissipates, the reader clicks the dot that pulses like a heartbeat, and the poem continues. This time, "Zoology" froze at the end of the first stanza. When I clicked for the next line, the program glitched: half of the dots clustered, half spread into the grid of their next configuration. The music stuck in a loop, and no further text appeared.

I decided to press ahead with my original lesson plan. What better object lesson could there be to illustrate the stakes of electronic literature than a poem that has become unreadable through some software glitch? Textual ephemerality and platform obsolescence have been defining issues for the study of electronic literature, as works increasingly become inaccessible, viewable only with floppy discs, CD-ROMs, specific Internet browsers, or outdated operating systems. Even Robert Coover, who predicted in "The End of Books" (one of a number of 1990s essays proclaiming the impending death of the book as a casualty of electronic textuality) that hypertext would make print literature obsolete, worried about the instability of hypertext authorship software.[1] A number of scholars and writers have intervened in the field by devising spaces and methods for archiving early electronic literature, with approaches ranging from the preservation of hardware to the construction of emulation systems. A list of just a few examples demonstrates the centrality of highly regarded scholars and academic institutions to the effort to thwart literary bit decay. For instance, the Deena Larsen Collection at the Maryland Institute for Technology in the Humanities (directed by Matthew Kirschenbaum) established a model for archiving the materials of a writer of born-digital texts; Nick Montfort's Trope Tank at MIT allows researchers access to games, electronic literature, and hardware, including the Apple II3 and Commodore 64; Lori Emerson's Media Archaeology Lab at the University of Colorado at Boulder preserves games and interactive fiction as well as technology ranging from 1970s personal computers to magic lanterns; Dene Grigar and Stuart Moulthrop's Pathfinders project, supported by a National Endowment for the Humanities grant, publishes video recordings of authors navigating their own electronic texts. All of these projects partake in what Moulthrop and Grigar have called "the attempt to preserve fragile artistic achievements against the eroding force of obsolescence."[2]

Ephemerality is the core condition of digital text. As Christine Borgman writes, "digital data are . . . more fragile than physical sources of evidence that have survived for centuries. Unlike paper, papyri, and paintings, digital data cannot be interpreted without the technical apparatus used to create them. . . . Unless deliberate investments are made to curate data for future use, most will quickly fade away."[3] Phenomenologically, too, the malleability and impermanence of text read on a screen makes digital textuality *seem* ephemeral, however durable the associated inscriptions may be on the surfaces of hard drives.[4] N. Katherine Hayles's influential coinage "flickering signifier" conveys this instability of text on screen, computationally rendered as an impermanent image:

> Marks on a screen . . . are the visible, tangible results of cod-ing instructions executed by the machine in a series of inter-related processes, from a high-level programming language like Java all the way down to assembly language and binary code. I hoped to convey this processural quality by the gerund "flickering," to distinguish the screenic image from the flat durable mark of print or the blast of air associated with oral speech. . . . The screen image is deeply layered rather than flat, constantly replenished rather than durable, and highly mutable.[5]

Flux lies at the heart of digital textuality. Ephemerality morphs into inaccessibility when a would-be reader cannot view those flickering signifiers because the required systems are no longer produced, work-ing, or readily available. Digital texts have troublingly short shelf lives.

Planned obsolescence is a guiding principle in the technology industry. "No future" has become an unstated design principle.[6] For instance, in the first decade after the iPhone was launched in 2007, Apple released fifteen versions—more than one every year. The differences are both aesthetic and technological: "The obsolescence introduced simply by changing the outward physical look of the device has proved a powerful technique to stimulate purchasing. . . . Beyond making the phone look different, the newer operating systems . . . are not designed with the old models in mind. . . . The company benefits from the need for consumers to purchase a new device and discard an old one (which we do at a pace of 426,000 discarded mobile devices *a day* in the United States)."[7] From the perspective of Joseph Schumpeter's theory of creative destruction, goods that are long-lasting make it difficult for companies to continuously increase sales and profits.[8] When products are designed for short-term use, businesses can "stimulate revenues . . . reduce competition . . . and increase prices for the replacement product."[9] Platform instability is driven by market forces. Apple projects the iPhone (and its own continued corporate existence) into the future not by producing models that will survive many years' use but by promising a constant cycle of newer versions available for purchase.

How, then, should one assess the durability of the print book? Medialogically speaking, the book has been unusually tenacious. While formats, printing technologies, and types of paper have changed, the appearance and organization of books has remained relatively stable. Today's readers may not be well versed in the conventions of incunabula, but they will recognize the Gutenberg Bible as a book, and its text remains legible after half a millennium. (By way of comparison, I own a CD-ROM of Michael Joyce's seminal hypertext novel *afternoon: a story,* first published in 1987. It is not compatible with either the hardware or software of any computer I own.) To grapple with the information cultures of the early 1900s and 2000s via the novel is to grapple with the scale of information, but it is also to ask how and whether information of any magnitude can persist and how such persistence might be linked to the novel's own archival project. The book is not just an agent of information preservation, a medium that preserves the data it stores; it is also an object to be preserved.

I have argued in the preceding chapters that descriptions of contemporary culture as post-print have their genesis in the modernist period, as similar discourse questioning the future of the book circulated widely. The book has survived declarations of its imminent death for well over a century, but competition from other media has intensified in the last several decades. The novel has frequently been haunted by scenes of the destruction of books and libraries. At the same time, a century (and more) of claims about the death of the novel have cast doubt on that genre's fate. These claims, too, are tied to media transitions—to the rise of cinema, television, and the Internet, among other cultural forces—as well as to the ongoing debate over whether and how the novel should change with the times (and whether, if it does change, readers will recognize it as such). This chapter examines the ways in which the death of the book has been forecast, the qualities that have enabled the book's persistence, and the oscillation between innovation and obsolescence that has been a driving force for new developments in information media and novelistic form alike. I argue that the novel has served as a testing ground for analyzing the consequences of the book's longevity; always already poised on the brink of its own obsolescence, it illuminates the different conceptions of futurity that have been bound up with the discourse of the death of the book.

LITERARY PLATFORM STABILITY

How might we theorize platform stability as an issue affecting not only born-digital literature but literature more broadly? As I have argued throughout this book, the metamedial novel's engagement with media and information culture during the early twentieth and twenty-first centuries has been conditioned by the novel's long-standing archival vocation. To regard the novel as an archive (or, in its more modern articulation, as an information storage medium) is to imagine a future time of reading. In other words, the novel's archival drive shapes the genre's thinking about its own perseverance (or obsolescence) as a genre, the perseverance (or obsolescence) of the media that store its narrative, and futurity itself. The future of the novel is not inevitably tied to the print book. Today, novels circulate as e-books, and some, such as Iain Pears's elegant time-travel hypertext novel *Arcadia* (2015), are designed to be read on digital platforms.[10] Yet the novel's continued existence has long been intertwined with that of the media format that most commonly stores it.

Many critical accounts of the effects of media transition treat the death of the book and the death of the novel as interchangeable and mutually determined. This is true of accounts that mourn these deaths, those that celebrate the apparent demise as clearing the way for new narrative forms, and more balanced analyses of what these shifts within the media ecology have entailed. The often-unremarked-on slippage between medium and genre is attributable both to the perception (discussed in chapter 2) that novel reading is an exemplary form of deep reading, grounded in the cultures of print and the book, and to the fact that the two largest media competitors to reading in the first two-thirds of the twentieth century, film and television, are terms that denote media as well as the content they carry. In the 1990s—the decade in which, Allison Muri has argued, "the book's death . . . began to register as . . . [something] other than provocative musings"—electronic media became serious competitors to the book as platforms for information storage and literary texts.[11] There are, of course, other causes to which theorists of the novel have attributed the genre's death, whether by this phrase they mean the gradual decline of interest in realism in the 1950s and 1960s; the passing of what Jonathan Arac calls "the Age of the Novel" (roughly 1830 to 1950), wherein the novel played a vital role in the construction of the national imaginary; or the broader Lukácsian argument that modernity exceeds the representational capacity of the novel form.[12] Even these accounts, however, invoke the close relationship between novel and book, whether by using the two terms interchangeably without serious reflection or by acknowledging media cultures' part in these shifts.[13]

As such, the book's persistence and stability have shaped ideas regarding the novel's future. In the early twentieth century, cinema marked the origin point for many of these "this-will-kill-that" narratives, as Arac terms the accounts of the novel's death due to media competition.[14] A century later, the assertion that books are dead has become so pervasive as to be played out; we are as certain that we are post-print as we are amused by the hyperbole of outdated claims such as Coover's argument that hypertext novels would become the future of fiction. Meanwhile, while novels have lost the privileged cultural role they once occupied, they still sell. "Proclaiming the death of the novel has become something of an apotropaic ritual," Margaret-Anne Hutton notes, even as novelists have become increasingly invested in what Jessica Pressman calls the aesthetic of bookishness.[15] Here I highlight three aspects of the book's endurance that are relevant to the history and future of the novel, particularly as its archival project intersects with its medial awareness: the continuation of the medium across time, the durability of individual books, and the ways in which the future of literature (and futurity more broadly) has been codified as a function of the continuation of books.

As a recognizably stable media object, the book encompasses a number of elements. As Nicole Howard writes, "a broad subset of technologies" has been refined over the history of the book, including "the creation of illustrations, the mixing of inks, the preparation of parchment, the making of paper, the casting of type, and the engineering of print."[16] As a format, the print book comprises three central components: the codex as the organizing form, paper as the primary support, and print as the inscription technology. The oldest extant parchment codex is a Roman artifact dating to 100 CE, although it took several more centuries for codices to become the dominant textual storage technology (and although scrolls coexisted with the newer technology through the fifth century CE). Paper may date as far back as the first century BCE, although many histories of the medium date the invention of paper later, to China in 105 CE.[17] At less than six centuries old, print is comparably young: it emerged in Europe with Gutenberg in 1450, with earlier antecedents in Asia.[18] Materially, today's print book differs in the production of these elements (wood-pulp paper, digital typesetting, and so on). Yet one reason that the book has achieved lasting dominance is that its three main components have been updated rather than exchanged or discarded. We still read text printed on surfaces that are light, flexible, and comparably inexpensive to produce; we still read long texts bound into books.

Conventions for how books are organized have also survived for centuries. A number of today's navigational aids were introduced in the thirteenth century, including the use of subject indexes and the division of books into chapters. By the fifteenth century, publishers had arrived at what Peter Stallybrass calls "the culmination of the invention of the navigable book," including features such as tables of contents and consistent pagination.[19] Other conventions, such as the labeling of the outside of a work with the name of its author and an indication of its contents, date back to scroll culture.[20] Robert Darnton notes the longevity of the book form to argue that media ecologies can flexibly accommodate both old and new media: "The staying power of the old-fashioned codex illustrates a general principle in the history of communication: one medium does not displace another, at least not in the short run."[21] As bookish structures such as the page find new expressions in digital platforms, readers continue to purchase, and to prefer, the familiar and very old form of the book.[22]

Individual copies of books have lasted because of their resistance to damage. As Amaranth Borsuk describes, "the codex is . . . a wonderful archival medium. It requires no software updates, can hold up in hot and cold climates, and, if printed and bound with quality acid-free materials, can withstand the oil of readers' hands, the jostling of being taken up and put back down, and numerous openings and closings that gradually break its spine."[23] One of the selling points of microfilm was its destructibility, making it advantageous for use in espionage. According to Vernon Tate, an executive secretary of the National Microfilm Association, "books may not be blown to bits or easily burned by fire," whereas "microfilms if capture is inevitable can be rapidly and completely consumed."[24] Comparisons of print books with digital texts hinge on similar contrasts. Deeply troubled by the practice of disassembling print books to scan them, Nicholson Baker argues that digital scans, like microfilm, are poor substitutes for books not only because they transform books from physical objects into images of their contents but also because they are more likely to perish over the course of time: "Digital storage, with its eternally morphing and data-orphaning formats, . . . is not now . . . an accepted archival-storage medium. A true archive must be able to tolerate years of relative inattention." "If you put some books and papers in a locked storage closet and come back fifteen years later, the documents will be readable without the typesetting systems and printing presses and binding machines that produced them," he argues, whereas "if you lock up computer media for the same interval . . . , the documents they hold will be extremely difficult to reconstitute."[25] Baker's statements are, to a degree, problematic: digital formats such as PDF and TIFF are considered archival standards, and his characterizations of library and archival practices have been roundly criticized by professionals in these fields. The emphasis he places on the stability and durability of books, however, remains a point of contrast in discussions of the ephemerality of digital texts. Little wonder that so many examples of the art practice that Garrett Stewart terms *bookwork* use books as the building blocks of architectural objects (hallways, spiral rooms, staircases); the appeal is not only the brick-like shape but also the book's physical and temporal stability.[26]

Yet books are by no means impervious to harm. Marcel Duchamp's *Unhappy Readymade* (1919) highlights the damage books suffer when exposed to the elements. Duchamp instructed his sister to open a geometry textbook, suspend it in the air, and leave it to endure rain, wind, and sun. The readymade meditates on "the changes which time effects, its proclivity for corroding, destroying, reducing to rubbish all that man builds," focalized through the damaged body of a book.[27]

Duchamp's title at once contextualizes books among his other ready-made objects as mass-produced commodities stripped of use value by their artistic resituation and anthropomorphizes them as responding emotionally to their own ruin. The term *biblioclasm*, meaning "the destruction of books," registers how cataclysmically such damage may be felt by readers. Tate argued that books (unlike microfilm) were too large to be quickly and completely consumed by fire in the event that they were about to fall into the possession of enemy forces, but book burnings remain one of history's tragic constants. Even where books do persist, bibliophiles cherish the markers of their decay—the yellowed pages, the "book smell" that contains an olfactory record of the paper's decomposition.[28]

In 1880, the printer and book collector William Blades published his treatise *The Enemies of Books,* listing fire and water as the two greatest antagonists. Fire and water are responsible for two of the best-known literary examples of biblioclasm: Ray Bradbury's *Fahrenheit 451* and William Shakespeare's *The Tempest,* which ends with Prospero's promise to renounce his power by drowning his book. (Blades's list of enemies also includes dust, ignorance, vermin, and now-antiquated concerns such as the effects of gas and servants.)[29] A UNESCO report on the destruction of libraries and archives in the twentieth century bears out Blades's characterization of fire and water as prominent causes of biblioclasm. The report catalogues more than 130 separate incidents and many millions of lost books. A number of these incidents were due to fires, set accidentally or intentionally. (The signifiers printed in a burning book also flicker.) Others account for books "drowned" via watermain breaks, serious flooding, and vandalism. The twentieth century also witnessed incalculable amounts of destruction through a force unanticipated by Blades: bombings. During World War II, according to UNESCO estimates, a third of all books in German libraries were lost.[30] As Ted Nelson put it in his advocation for digital archives, "books disappear. Knowledge of the past is lost. Libraries *burn,* and each time, we are diminished."[31]

Books, then, are ambivalent media. Their long-standing stability makes them perpetually old media even as their robust propensity to endure confers on them an ideal archival persistence. As information storage media, books are vehicles for the preservation of the past; as media whose platform stability and solid construction project them forward in time, they index futurity. At the same time, books have been haunted by the constant specter of their annihilation, whether through declarations of the death of the medium or specific incidents of biblioclasm. As a result, their future existence often functions as an assurance of futurity. For the book to persist, in this view, is for knowledge and culture as a whole to persevere. As Geoffrey Nunberg argues,

> when people talk about the future of the book they have some-
> thing else in mind: works of literature, belles lettres, scholar-
> ship, and criticism, as well as . . . journalism, reportage, and
> general informative writing. . . . Understood in this way, con-
> cerns about the future of the book are something more than
> reflexes of the nostalgia we feel for threatened artifacts like
> the steam locomotive or the pinball machine. "The book" here
> stands in a metonymy for all of the material circumstances of
> print culture—not just the artifacts it is inscribed in, but the
> forms and institutions that have shaped its use.[32]

In literature, scenes of the destruction of books frequently mark the limit of civilization, even of the Anthropocene. Many recent dystopian novels feature post-apocalyptic worlds in which books are either rare or nonexistent and the population is illiterate.[33] Humans may survive in a post-book world, such narratives assert, but humanistic culture cannot.

Novels are similarly entangled with the envisioning of the end of civilization as the end of the book. Discussing Nunberg, Kathleen Fitzpatrick slides from his "end of the book" to "the loss of the novel": "the loss of the novel is the loss of libraries, of literary studies; the death of the novel is the death of the traditional humanities."[34] The book is figured by and through the novel in post-civilization narratives. Narrative forms, moreover, are useful for theorizations of futurity; they allow readers to cast their minds into the possible futures, inhabiting the speculative. The novel has contended with the possibility of its own end for so long that it is uniquely suited to reflect on futurity. "The novel has been proclaimed dead or dying for nearly as long as it has been alive," writes Fitzpatrick.[35] The genre endures, but only through constant renewal. In all of these ways, novels have been influential to theorizations of futurity, particularly as that futurity is caught up in the media culture of the book.

The modernist moment offers both an origin point for contemporary (so-called) post-print culture and a counterpoint for conceptualizing the obsolescence and destruction of books. At times, books served as focal points for modernist anxieties regarding the perpetuation of culture, as they have done in recent fiction. Paul Saint-Amour has argued that, during the interwar years, a number of "air war fantasies" depict the end of humanity following devastating warfare—he cites titles including *People of the Ruins* (1920), *The Collapse of Homo Sapiens* (1923), and *Armageddon 2419 A.D.* (1927)—demonstrating "a growing preoccupation with the effaceability of literacy, numeracy, and the archive."[36] As Saint-Amour recounts, H. G. Wells's *The War in the Air* (1908; reprinted in 1921) contains a scene in which two men find a roomful of books. When one picks up a book, it crumbles into dust.[37] The book is an emblem of the demolishing of civilization, obliterating even humanity's ability to represent and preserve its own history.

The physical precarity of the print book was newly palpable by Wells's time, as production methods introduced in the 1850s had resulted in books with shorter shelf lives. Earlier paper was made from rag; during shortages of cotton and linen, publishers turned to the cheaper alternative of wood pulp. This change, combined with the use of alum rosin sizing to treat the surface of paper, made for paper that was highly acidic and would therefore deteriorate at a much higher rate. Consequently, "the period from 1850 to . . . [the end of the twentieth century] has often been considered 'the era of bad paper.'"[38] Recall from chapter 3 Orlando's description of modern books as "ephemeral," "[seemingly] bound in cardboard and printed on tissue paper."[39] Texts written on vellum have survived for more than a millennium, and much handmade paper produced before the 1850s remains in "very useable condition" three hundred years after it was produced. Books produced during Woolf's lifetime, in contrast, were expected to last less than half a century, a span shorter than Woolf's own life.[40] Books that had been produced in the second half of the nineteenth century had begun to deteriorate by the start of the twentieth, making obvious the fate that books printed during the modernist period would face. In a collection of essays advancing "The Case for Books," Darnton casts the book as a far superior storage medium compared to digital media, but he does so in terms that stress the lessened longevity of the modern book: whereas "all texts 'born digital' belong to an endangered species, . . . nothing preserves texts better than ink imbedded in paper, *especially paper manufactured before the nineteenth century,* except texts written in parchment or engraved on stone." Thus, "the best preservation system ever invented" is not the book generally but specifically "the old-fashioned, pre-modern book"—an assessment paradoxically making the case for books' continued relevance by focusing on those that are artifacts of a bygone age.[41]

Because of books' connections to earlier historical periods, accounts of the end of books were less fraught with anxiety during the modernist era. In many contexts, books were dismissed with the pragmatic desire to modernize textual media or were castigated with revolutionary zeal. As I discussed in chapter 3, microform advocates echoed Bob Brown's complaint that readers were receiving "twentieth century reading matter in fifteenth century book form."[42] In this view, the book, as an antimodern form, denoted the stagnation of culture. This is the case when *Ulysses*'s Stephen Dedalus, feeling oppressed by the weight of literary history in its physical materialization in Ireland's National Library, thinks of the shelved books as "coffined thoughts in mummycases." Dedalus and Leopold Bloom prefer to inscribe their texts in ephemeral media. Bloom traces letters in the sand at the beach, aware that they will be "washed away," while Stephen imagines writing his "epiphanies . . . on green oval leaves."[43] The apotheosis of the idea that books were sorely obsolete arrived in Marinetti's call in the Futurist Manifesto to "Go ahead! Set fire to the shelves of the libraries!"—a cry that would find its eerie echo in the Nazi book burnings of the 1930s.[44] Even those modernist practitioners who did not call for the end of the book but were instead invested in its aesthetic potential—an investment we see in writers' experimentation with the visual properties of the page, in the work of book designers and typographers, and in the burgeoning of artists' books—sought to transform the book, driven by the conviction that the medium could not simply continue in its traditional form.

In a 1927 radio broadcast co-written with Leonard Woolf, Virginia Woolf imagined a future in which books would be disposable by design: "It is absurd to print every book as if it were fated to last a hundred years. The life of the average book is perhaps three months. Why not face this fact? Why not print the first edition on some perishable material which would crumble to a little heap of perfectly clean dust in about six months time? . . . Thus by far the greater number of books would die a natural death in three months or so."[45] Woolf presents a modernist version of designed obsolescence in keeping with the growth of disposable consumer commodities. As a response to the subject of the radio debate ("are too many books written and published?"), her suggestion for disposable books directly addresses the intertwined issues of information scale and the proliferation of textual media. She envisions books crumbling into dust, not due to war or the passage of centuries but as a solution to the problem that chapter 1 explored: what to do when there is too much to read, there are too many books published, there is too much information.

Melba Cuddy-Keane cautions that, "given the staged nature of the broadcast, we should be wary of assuming that its arguments transparently represent each speaker's views."[46] We need not, however, assume that Woolf is offering an entirely serious or practical plan here in order to recognize that her solution to the overproliferation of books draws on the fact that so many of the period's textual media were designed to be ephemeral. She aligns books with more modern and ephemeral textual media—media such as newspapers, discardable once consumed. Woolf positions the print book firmly within the realm of modern information culture. In Walter Benjamin's description, "the value of information does not survive the moment in which it was new. It lives only at that moment."[47] Woolf pictures the transformation of books into a medium whose rate of deterioration would keep pace with that of early twentieth-century information, living only for a brief moment.

UNREADABLE NOVELS, UNREADABLE BOOKS: INNOVATION IN MEDIA AND THE NOVEL

As I have shown, the scale and management of information in the early twentieth and twenty-first centuries have had a significant impact on novelistic experimentation, and this experimentation offers insight into modern information mediation. An implicit through line accompanying this argument is that novelistic engagement with information has hinged on what it means to read—and, less intuitively, on what it means for a novel or an informational text to be *unreadable*. As metamedial experimentation has drawn attention to how print books mediate information (in contradistinction to the mediations of interface-based media and information systems), this experimentation has had the effect of foregrounding the materiality of textual media. Metamedial novels emphasize how the mediality of the print book produces tension between its role as an information vehicle and its role as a material object. In the process, novelists examine the conditions under which texts become unreadable.

I am discussing unreadability in terms quite different from those of poststructuralism, whose concept of unreadability hinges on the plurality of possible interpretations. In contrast, I am attending to literary texts' unreadable qualities as they are linked to the unreadability of textual storage media. Media that store information as text raise implications that are different from those of media that record non-humanly-readable inscriptions or other data traces. Phonography, for instance, was a form of nineteenth-century information storage that was "illegible" to humans in the sense that, as Lisa Gitelman describes, "only machines could 'read' (that is, 'play') those delicately incised grooves."[48] The consequences are vastly different for a form such as a novel or a medium such as a book, whose human readability is a basic assumption. Metamedial novels have explored the conditions under which textual information that should be humanly readable is rendered illegible and how these conditions then challenge the archival project of the novel. Here, I am focusing on unreadability as it arises when texts can no longer be accessed, whether because a resource such as an individual book has been destroyed or because an entire reading platform has obsolesced.

As we have seen, another category of unreadability results from the overabundance of textual content. The complaint of "too much to read" examined in chapter 1, like its contemporary analogue "TL;DR," asserts the inability to read due to the overwhelming proliferation of material. In this situation, the issue is not that any single text is actually illegible. It is, rather, the belief that the scale of textual culture has become overwhelming. The quantity of available material makes it difficult for the reader to decide which texts to select; at the same time, the archive of available literature becomes a corpus that can be managed (navigated, skimmed, searched, datamined) only via the mediations of computers or other automated systems. As I have shown, a major novelistic response to this condition has been to reproduce the experience of information scale's cognitive overload within the limited space of the print book, using the arrangement and structure of the text to stymie conventional reading practices. Such novels highlight the psychological impact of challengingly large stores of data and of the reading practices they necessitate or preclude.

It is no coincidence that the critique of experimental literature as unreadable arose during the modernist period nor that this critique has resurfaced regarding contemporary avant-garde writing. Consider Shane Leslie's summation of *Ulysses* as "unreadable and unquotable, and . . . unreviewable."[49] Those skeptical of the literary merits of modernist writing pilloried it as unreadable primarily because of its difficulty. As T. S. Eliot famously declared, "poets in our civilization, as it exists at present, must be *difficult*."[50] A literary text may be challenging without being lambasted as unreadable, and simple, straightforward works were not immune from this charge. (Eliot derided Robert Frost's poetry as not only "uninteresting" but also "unreadable.")[51] Yet the common thread in dismissals of experimental literature on the grounds of unreadability is that such literature is opaque—indistinguishable from hoaxes or gibberish—because, as Leonard Diepeveen writes, the "difficult work is excessive," "overload[ing] viewers and readers."[52]

Like information overload, the difficult, unreadable, experimental text taxes its reader by presenting too many details to easily navigate or assimilate and by lacking a recognizable structure for doing so. As Diepeveen notes, this critique was often grounded in the problematic contentions "that art should not require a conscious (or unconscious) employment of facts" and that factual knowledge (information, although Diepeveen does not use the term) was mixed into the literary work arbitrarily, without regard for organizing principles: "Skeptical readers described this kind of knowledge as isolated, disconnected bits of stuff; it wasn't *real*, integrated knowledge, but an idiosyncratic, haphazard collection. A standard trope for skeptics reviewing modern difficulty was to list all the sources used by the writer and to argue that what these sources resulted in was just that—a list, and a poorly organized one at that."[53] Put another way, as modernist writers imitated the formal structures of information media, difficult modernist literature was described in terms similar to information shock.

Chapters 3 and 4 examined a second condition of unreadability: that of the nontextual, material data collected by a medium via its material interactions with readers' bodies and the physical world. This condition encompasses all the ways in which textual media store and accumulate nonlinguistic information in addition to the text they contain, whether this be in the form of photographs or other images, the haptic traces left by readers, or the bibliographic codes of the print book. As the novels I have examined demonstrate, the print book's tactility makes it an ideal medium for emphasizing such unreadable, haptic information. This tactile quality distinguishes the book from newer, interface-based textual media, especially as regards the history of reading interfaces (whether the screens of microform readers or mobile phones). Interface-based reading, as I have argued, has been the product of new techniques for storing ever-larger quantities of data in ever-smaller amounts of physical space. What novelists from Woolf to Foer have contended, however, is that no quantity of legible information can comprise an adequate substitute for the living, embodied presence of the human subject—a presence the book's tactility mimetically reproduces.

To the critique of experimental literature as unreadable due to its density of narrative information, metamedial experimentation adds the perception that a text need not be read because its material instantiation is more important or interesting than the text it contains. This sense is adjacent to, but distinct from, the principle invoked when conceptual literature is described as unreadable. "The new writing might be best not read at all," argues Kenneth Goldsmith; "it might be better to think about."[54] That ethos (or critique, depending on whether one asks conceptual writers or their critics) assumes that the point of the literary text is the concept it embodies rather than the language the author has crafted to represent this concept. With metamedial literature, the notion that the text can be disregarded stems from the opposite assumption: that reading the text is not necessary because the text has nothing of substance to communicate. Instead, this argument asserts, the author has sacrificed literary content to material innovation.

This view, more likely to be disseminated by the reading public than by academics, but still pervasive, disparages innovation with the book (or other literary media) as a superficial gimmick. Consider this review of Mark Z. Danielewski's *The Familiar, Vol. 1* in the *Guardian:* "The author of *House of Leaves* has published another novel full of typographical high jinks—but the rest of the novel is inscrutable, if not indecipherable. . . . The problem with *The Familiar* isn't that it's difficult; it's that it's unreadable. Take away the typographical gimmicks, the frequently unfathomable dialogue, and the confusing storylines that pass for a plot, and you're essentially left with nothing."[55] The reviewer's central complaint—that "a novel that appeals chiefly to people who like to look at books rather than read them isn't a meaningful contribution to the world of literature"—takes for granted that reading a book and looking at a book (a phrase for which we might more generously substitute "appreciating the aesthetics of print") are mutually exclusive. *The Familiar* is certainly a challenging novel: its reader must track nine interwoven plot lines and parse prose that ranges from Singlish dialect to the idiolect of a precocious twelve-year-old. The reviewer, however, grapples with none of this narrative or linguistic difficulty, instead dismissing the novel on the basis that its visual and material properties must necessarily gloss over a dearth of worthwhile content.

I am dwelling on the critique of unreadable-because-materially-gimmicky because it demonstrates how innovation in the metamedial novel has been tied to broader thinking about media as well as how responses to novelistic innovation in this vein differ from responses to innovation with technology. The contention that metamedial novels are conceptually and literarily insubstantial because of their mediality is deeply flawed. As I have argued throughout this book, metamedial novels engage with the book's form as a means of engaging with questions about the limits of literary representation, the vocation and cultural work of the novel, the limits and viability of the print book, and the nature of meaning making in an information ecology where such processes are increasingly ceded to nonhuman agents. Metamedial novels analyze how literary texts mediate ideas even as they navigate the textual media that mediate them. Yet literary experiments with media have repeatedly been disparaged on the grounds of style over substance.

Take the case of early hypertext novels. Reif Larsen, who describes his metamedial print novel *The Selected Works of T. S. Spivet* (2009) as "essentially an exploded hypertext," described his initial reactions to digital hypertext literature in terms that echo the critique that metamedial literature is gimmicky: "Hypertext fiction, with very few exceptions, was mostly bells and whistles."[56] The phrase "bells and whistles," which likely derives from fairground organs, gained prominence in the late 1970s and early 1980s in the context of computing. (For instance, the *Oxford English Dictionary* cites an example from the *Sunday Times* in 1984: "There are more than 600 microsystems on the market so it is hardly surprising that the manufacturers have taken to hanging a few bells and whistles onto their machines to get them noticed.")[57] Whereas the phrase surfaces commonly in criticisms of electronic literature and print metamedial literature, in the context of computing it usually signifies extra features. These features may be inessential, but their existence is rarely taken as indicative of the system's overall uselessness. Given the demands of neoliberal capitalism, the technology industry must continuously innovate in order to thrive. Novelistic innovation, in contrast, runs the risk of seeming superfluous.

The past two decades of genre theory should make one wary of the reifications of "the novel" that underlie such dismissals. As we know, the novel has been defined via its indefinability, its volatility as a form that constantly seeks renewal. The realist novel may be the longest-standing and the central template for the genre, but the novel has cycled through different subgenres throughout its centuries of existence. In his distant reading study of British novels published between 1740 and 1900, Franco Moretti argues that "the novel does not develop as a single entity . . . but by periodically generating a whole set of a genres, and then another, and another [. . .] Both synchronically and diachronically, in other words, the novel is *the system of its genres*." According to Moretti, "all great theories of the novel have precisely reduced the novel to one basic form only (realism, the dialogic, romance, meta-novels)," an action that has had the effect of "eras[ing] nine tenths of literary history."[58] Jonathan Culler defines genre differently—as "norms in the process of reading," "set[s] of instructions about the type of coherence one is to look for" in a given genre—but, like Moretti, he expresses concern about literary history's neglect of outlier texts. Culler contends that the works he terms "nongenre literature" are "unreadable" in the sense that they fall outside of readers' genre expectations. A genre may undergo "radical changes of convention" (for instance, in the shift from social to psychological realism) while still operating under the same laws of readability. In contrast, writerly, resistant works (such as Stéphane Mallarmé's *Un Coup De Dés* and Joyce's *Finnegans Wake*) "must be read at a metaliterary level: the level at which the acts of reading and writing are posed as problematic." Far from being outliers, Culler argues, such examples of nongenre literature are "central to the contemporary experience of literature."[59] Metamedial novels may present unusual forms of literary innovation in their defamiliarization of the reading medium, but their experimental tactics are in keeping with contemporary fiction and the history of the novel.

Metamedial novels demonstrate the interconnections between genre stability (anxiety about the future of the novel and interest in the genre's potential reinvigoration through formal innovation) and platform stability (anxiety about the persistence of the print book and interest in the potential of renewed interest in the medium's affordances to stave off its obsolescence). As the "bells and whistles" charge suggests, these novels recall the reader's attention to the material presence of books—their nature as objects and things as well as texts to be read. This is neither new nor exceptional: book historians and literary scholars such as Leah Price and Patrick Collier have documented that books' materiality has mattered throughout literary history.[60] In their explorations of unreadability, however, metamedial novels go beyond emphasizing their status as material objects; they also stress that they are material *media*—that the conveyance of meaning is inseparable from the embodiment of text. In other words, by showing how unreadability arises from different medialogical conditions, these novels dismantle the words-things binary.

Novels that anticipate or perform their own destruction center the fact that the book, as a medium, bridges material objectness with textual transmission. In so doing, they interrogate the processes of media obsolescence.[61] Gretchen Henderson's print hypertext *Galerie de Difformité*—part novel, part museum, part treatise on aesthetics— draws on the theory of Jerome McGann and Lisa Samuels, inviting its readers to "deform" it.[62] Readers are asked to fill in blanks, draw self-portraits, and otherwise personalize the text. In a section titled "The Destruction Room," Henderson suggests that readers transform the entire book into any number of new objects, from origami cranes to confetti.[63] While foregrounding the book's position in an increasingly non-print-centric media ecology (many pages display QR codes linking to digital content), *Galerie* also rehabilitates the book, casting its deformability as a sign of malleability rather than vulnerability. Henderson also positions *Galerie* as the latest in a long line of bookish innovations, from the marbled page of Laurence Sterne's *Tristram Shandy* (which forms the background image for *Galerie's* cover) to surrealist and Dada experiments with the book. The very destructibility of the book, Henderson reminds us, might grant the medium rich afterlives.

To take a modernist example, John Lurz recounts how the editions of *Ulysses* he owns have fallen apart over the course of his readings and rereadings: "By the time I reached Molly's final 'Yes,' the pages had come unglued from the paperback spine and fell out in large chunks every time I opened the book. . . . The disintegrated book stood on my shelf as an embodied record of the weeks I'd spent on my reading."[64] Lurz considers this disintegration to be a serendipitous echo of the novel's typographic meditation on the embodied nature of the act of reading: for him, *Ulysses* dwells on the corporeality of the book's typographical body in order to force the reader to consider how her own body is implicated in the processes of reading the text. It also, he writes, forces the reader to acknowledge corporeal mortality. I add that Joyce's reflections on the condition of the books in Bloom's bookshelves, featuring covers that have faded or been lost, "obliterated" titles, and missing pages, anticipate his own book's fate—if not the fragility of the later, mass-produced editions owned by readers like Lurz and myself, at least that of the edition he oversaw so meticulously, whose faded and brittle blue wrappers must have given pause to many researchers besides myself as their hands, archivally gloved, hesitated over the pages.[65] Printed books, Joyce reminds us in *Ulysses,* are likely to lose their integrity over time, perhaps eventually joining the waste-paper economy of Dublin's other printed throwaways.

Similar issues of perishability and obsolescence recur throughout the texts I have examined in this book. Orlando's manuscript survives the centuries but in increasingly distressed form. Danielewski chose to publish the electronic edition of *Only Revolutions* via iBooks rather than as a standalone iOS app, due in part to his concern that the maintenance of an app presented greater challenges to the e-version's longevity.[66] At the time of this writing, *theMystery.doc* is only three years old, but already some of the online content archived in its pages is no longer available on the Internet.[67] Metamedial novels, finally, are records not only of the information they contain (whether pulled directly from real-world documents or invented descriptions of fictionalized worlds) but also of how storage media have endured, or ceased to endure, at different moments over the course of the past century. The novel, Daniel Punday has argued, is distinguished from media forms such as painting and music by its "facility at describing what is absent," an ability that makes the genre "particularly good at recognizing other media and thus at describing media limits"—including its own.[68] Always evolving and innovating yet always dodging declarations of its end, merging a generic investment in futurity with nuanced attention to the limits and thresholds of media, the meta-medial novel speaks with unique relevance to the processes by which information, texts, media, and even genres might become unreadable or otherwise lost.

KINDLING: PRINT AESTHETICS, E-READERS, AND THE FUTURE LIBRARY

I have avoided the impossible (if tempting) task of predicting the book's future. Research indicates that readers continue to prefer the print book as a platform for pleasure reading, especially in the case of longform genres such as the novel.[69] The degree to which the book's existence will be imperiled by newcomers to the media ecology (from increasingly haptic digital interfaces to the new regime of voice interface in the home) remains to be seen. What I have charted instead is a history of the book's future (or lack thereof), as it has been championed or dreaded at two significant moments of media transition. As we negotiate a world in which humanity's future is imperiled by forces such as climate change, literary studies has turned to studies of futurity as they have been articulated within queer studies, ecocriticism, and other fields. Critical scholarship has also increasingly looked backward, with the rise of failure studies and media archaeology. These approaches examine technologies that failed to realize the potential their inventors anticipated as well as those that were only ever speculative, analyzing what Reinhold Bauer calls the "extreme fragility of the innovation process" and asking how and why media and technologies fail.[70] As

Erkki Huhtamo and Jussi Parikka argue in their overview of media archaeology, "dead ends, losers, and inventions that never made it into a material product have important stories to tell."[71] These critical approaches counter the digital moment's rapid forgetting of old media by studying the factors that lead to failure and obsolescence and by memorializing old media through undertakings with evocative (and funereal) titles such as "The Dead Media Project."[72] Media archaeology is antiteleological: Eric Kluitenberg describes it as "a critique of progress" and "an alternative to the dominant writing of media history," Huhtamo and Parikka as "alternate histories of suppressed, neglected, and forgotten media that do not point teleologically to the present media cultural condition as their 'perfection.'"[73] Grounded in the past, these approaches reject a future-orientation.

Shelf Life

Throughout my analysis I have drawn on media archaeological approaches, seeking to enliven the dialogue between media archaeology and new media theory by juxtaposing accounts of old and new information media insofar as they have shaped the cultural landscape with which novelists have engaged. The history of textual information storage media, for instance, dovetails with the history of concerns about what constitutes socially sanctioned "good" reading, from late Victorian anxieties about the expansion of reading publics and print cultures, to modernist attempts to differentiate the difficult task of reading elite literature from the popular entertainment of mass culture and mass media, to National Endowment for the Arts studies of the rates of pleasure reading as books compete with digital media for readers' attention. The novel has focalized these histories, both in its narrative representations of books and reading as they have been affected by information culture and in its function as an emblem of a mode of reading taken to be synonymous with everything from deep attention to humanistic thought. It is this interpenetration of literary and media cultures that becomes most helpful for theorizing the future, even as it deepens analysis of the past. In her compelling analysis of texts and technologies, Gitelman writes that "media history and literary history share the same groundwater."[74] We might say the same about media futures and literary futures. They are co-imagined and co-constituted, and the discursive register of the speculative creates one of few available means for accessing and assessing possible futures.

I turn here to an examination of two different visions of the future of literary reading as it intersects with print culture. First, I look at the Future Library Project, which is in the process of assembling a new literary anthology scheduled to be published in the twenty-second century, thus tying the anthology's viability to the future of the print book (which the project strives to protect). Second, I explore how the aesthetics of print media might be transferred to a digital reading platform, the Amazon Kindle, and what consequences this transference might entail. I consider these issues by analyzing the Kindle's marketing material, Stephen King's novella *UR,* and Sebastian Schmieg and Silvio Lorusso's artist's book *56 Broken Kindle Screens.*

The Future Library Project forges connections to the future a century from now. Spearheaded by Katie Paterson, a Scottish artist whose works reflect on time and scale, and in conjunction with Norway's New Oslo Public Library, the Future Library will collect one unpublished book a year between 2014 and 2114, at which time the manuscripts will be published. Although it remains to be seen how heavily novels will feature in the final anthology, all of the writers who contributed books during the project's first five years of activity—Margaret Atwood, David Mitchell, Sjón, Elif Shafak, and Han Kang—have previously published novels, and the novel is the genre for which several are best known. At an annual ceremony, a new manuscript, printed on paper, is presented to the Future Library Trust. To ensure the availability of raw materials for the 2114 anthology, Paterson and her associates have planted a thousand spruce trees in the Nordmarka Forest near Oslo. The Future Library Project proposes to ensure the future of literary reading by ensuring the survival of the print book as well as the ecosystems that make print production possible.

The project is equal parts speculative and hopeful. The manuscripts will be kept in a special room in the New Oslo Public Library, which features the wood of the trees felled to clear space for the new forest. The room will be open to the public as a space of silent contemplation as the locked-up manuscripts flaunt their inaccessibility to their not-yet-readers. Atwood writes, "People will be able to go into that room and see the titles and the author and *imagine* what's in the boxes." The book she has written "is a letter to the future," she continues, characterizing the Future Library as "a very hopeful thing because . . . you're assuming that there will be people a hundred years from now . . . [and] you assume that the forest will grow . . . [and] that the library will still be there. You're assuming that people will still be able to read and that they will still be interested in reading."[75] Whether or not the project will meet its goals remains to be seen. Climate change is one of many threats to the Future Library's existence: "The next 96 years do not look promising for the seedlings, which are more vulnerable than their ancestors to all manner of man-made disasters: the storm surges, wildfires, heat waves, and droughts precipitated by global warming, as well as the less dramatic possibility that . . . people will simply stop tending to them and the books that are their fate."[76] Should the trees in the Nordmarka Forest survive disasters or neglect, they will cease to exist as trees when they are made into paper, an act of deforestation that mimics the eco-collapse that the library symbolically forestalls. The Future Library has much in common with other slow-time art interventions, but it is also a decidedly literary encounter with slowness and futurity. Paterson's only rule besides barring authors from disclosing the contents is that the manuscript be all in text—no images. Each manuscript is "a letter to the future" in the literal sense of being comprised of letters, of text. Paterson describes the library's texts as literary time capsules: "Looking back over 100 years, who could have predicted the sea changes that have taken place since 1914? I hope the writings in the Future Library Anthology will contain crystallized moments from this era to the next."[77]

To safeguard the future time of reading, Paterson turns to the past: the library's futurity is managed via the terms, materials, and institutions of print culture—books, paper manuscripts, wood pulp, libraries. In addition to growing trees and preserving manuscripts, the Future Library has preserved a printing press (and instructions for its use), in case that technology should have become unavailable by 2114. As Paul Benzon writes, "the external face of the *Future Library* is resolutely analog, rooted in the tangible, visible materiality of wood and paper. These old-media trappings are at once aesthetic and technological: the living forest, the bound paper copies, and the wooden interior of the library room that collectively compose the project seem to speak strongly on behalf of the analog as a format with the capacity to persist long after the obsolescence of any single given digital format."[78]

The Future Library challenges the assumptions about temporality and media transition that typically underpin discussions of literary unreadability. These texts are unreadable not because they have become inaccessible over time but because they have not yet have been made accessible. As Jeffrey Di Leo characterizes the work, "Paterson's Future Library is therefore more like a safety-deposit box or a vault in a bank than a library item or collection."[79] Although, as Benzon points out, each book added to the Future Library was presumably first composed as a digital file—and although Paterson has stated that manuscript submissions include a copy, presumably a digital file, that is "archived and encrypted"—the Future Library figures print as the primary medial support.[80] In this version of the future (of reading, of the novel, and of humankind), the digital becomes the residual medium.

A different view of the relationship between the print book and the future of literary reading has emerged with the Amazon Kindle. Long the most popular e-reader on the market, the Kindle has thrived on Amazon's promise to provide a reading experience rooted in the aesthetics of print culture but liberated from the shortcomings of the book. While its ability to store thousands of novels (or other texts) in a compact, lightweight frame is an advantage of digital storage, the

Kindle retains features bibliophiles prefer: a simplified, focused reading surface; non-eyestrain-inducing black text on a white field; and a streamlined form designed to invoke the canonical achievements of print culture alongside the ease of pleasure reading. The device uses "E Ink" and "Paperwhite" technology, names gesturing toward the comfort of the printed page as a reading support. While the Kindle recalls Brown's claim that readers of the future would be able to carry portable libraries wherever they went, he would certainly have decried its fidelity to the look and structure of the printed page. As Bonnie Mak has argued, however, the page is a "conceptual structure" that originated with scroll culture.[81] The Kindle thus represents not only the replacement of the printed book with the digital e-reader but also the ability of reading structures such as the page to persist across media forms. With the Kindle, Amazon presumes that the future of reading lies not with the book but with a digital reading platform permeated by an aesthetic of bookishness.

Amazon's design and marketing decisions also portray the Kindle as guaranteeing ecological futurity by mitigating the environmental impact of print book production. The logo depicts a young, boyish figure in silhouette reading a rectangular shape that might be either a book or a tablet, seated at the base of a tree whose branches sprawl overhead. It is difficult to calculate the relative environmental repercussions of book production and the manufacture of electronic reading devices. According to the Sierra Club, an average tree (perhaps the Platonic ideal of treeness evoked by the Kindle logo?) produces between 10,000 and 20,000 sheets of paper. In 2013, more than 20 million tons of paper were produced in the United States, requiring somewhere between 55 million and 100 million trees.[82] Wood pulp has been central to book production for nearly two centuries, and the etymological roots of the word *codex* reflect the form's dependence on living materials: the word derives from the Latin *cōdex,* meaning "trunk of a tree."[83] Novelists and literary scholars may conceive of the end of the book as the limit of the Anthropocene, but environmentalists have asked whether a world in which books have been replaced by e-readers will be more sustainable. According to Daniel Coleman and Gregory Norris, "with respect to fossil fuels, water use, and mineral consumption, the impact of one e-reader payback equals roughly 40 to 50 books. When it comes to global warming, though, it's 100 books; with human health consequences, it's somewhere in between."[84] Such comparisons presume that a reader will continue to use the same e-reader rather than upgrading every few years. Globally, e-waste creates substantial problems.[85] Meanwhile, many book publishers have used paper treated with alkaline buffers, which reduce the acidity and deterioration rates of wood-pulp paper, meaning that those forty or fifty books a single e-reader can replace will likely have a longer shelf life than the e-reader does.

Whatever the actual environmental implications of switching from books to digital platforms, e-readers such as the Amazon Kindle are marketed as technologies capable of combining a love for reading with a lessened environmental footprint. When I purchased my first Kindle in 2010, the device arrived packaged in brown recycled cardboard, proudly touting its environmentalist bona fides. The screen displayed a static image: the Kindle logo, beneath which was a definition for *kindle*: "/kindl/ [*v. trans*] Light or set fire. Arouse or inspire (an emotion or feeling): *a love of art was kindled in me.* [*intrans*] (of an emotion) be aroused: *She pressed on, enthusiasm kindling within her.* [*intrans*] become impassioned or excited: *the young man kindled at once.*" The name Kindle, as Nicholson Baker points out, is "cute and sinister at the same time," suggesting a Bradbury-esque fate for the device's print antecedents.[86] From the recycled cardboard to the echo of kindling, the design underscores the message that this digital platform allows the reader to enjoy the pleasures of print reading—the haptic erotics of "becom[ing] impassioned"—while better caring for the environment (no more dead trees!).

Although Amazon portrays the Kindle as a digital device that will safeguard the survival of print's aesthetics, Stephen King's novella

UR recasts it as antithetical to the book: as a reading platform whose digitality and futurity are at once inextricable and threatening. *UR* was commissioned by Amazon and initially published exclusively as a Kindle e-book, with its release timed to coincide with that of the second-generation Kindle in February 2009. (In 2000, King had pub- lished *Riding the Bullet* as a digital download, becoming "the first major best-selling author to release a book exclusively in electronic format.")[87] In the novella, Wesley Smith, an English instructor and aspiring novelist, orders a Kindle. Pre-Kindle Smith is a relic of print culture. His name recalls elite East Coast liberal arts colleges, and he carries a briefcase containing serious literature (Roberto Bolaño's maximalist novel *2666*) and "a bound notebook with beautiful mar- bleized boards."[88] Despite his pretentions and aspirations, he teaches at an institution that King characterizes as an obscure and mediocre regional university, and he has failed to write his own novel. Despising digital technology, he epitomizes the failure of traditional humanistic literary culture.

Smith orders a Kindle to prove himself capable of changing with the times, but the device he receives decenters his temporality. This Kindle grants access to a near-infinitude of potential libraries as well as to the future. The device that arrives is improbably pink, a color at once pointedly anti-utilitarian, disruptive of Amazon's muted bookish aesthetics, and a sign of its genesis outside of Smith's reality. After marveling at the Kindle in passages that teeter tediously close to Amazon advertising copy, Smith discovers that the menu options include experimental features labeled "UR FUNCTIONS" (loc. 404). Each Ur is an alternate universe; by typing random numbers, he can access novels written by writers beyond from his own reality: "He had searched for Ernest Hemingway in two dozen of the Kindle's almost ten and a half million Urs, and had come up with at least twenty novels he had never heard of. In one of the Urs . . . , Hemingway appeared to have been a crime writer. . . . He *always* wrote *A Farewell to Arms,* it seemed; other titles came and went, but *A Farewell to Arms* was always there and *The Old Man and the Sea* was *usually* there" (loc. 589). (Smith himself remains an "unpublished loser" in every universe [loc. 625].) King's fictional conceit parodically distorts the Kindle's main selling point into a Borgesian fantasy: that this platform, combining digital textual storage with the ability to download hundreds of thousands of books from Amazon's online store, replaces the one-to-one book-to-novel relationship with an infinite library stored in a single device.

From *UR*'s beginning, King aligns the Kindle with literary futurity, portraying it as "new school" electronic reading that directly opposes Smith's decidedly "old school" preference for print (loc. 1674). This association increases when Smith finds himself able to access the literary histories of alternate realities: "He thought of all the writers whose passing he had mourned . . . ; one after another, Thanatos stilled the magic voices and they spoke no more. But now they could" (loc. 625). This Kindle defies death by making pastness an irrelevant state. The alternative pasts of writers such as Hemingway become future libraries for Smith. The title hints at the device's confounding of temporal linearity: *ur* denotes a primitive state or the earliest incarnation. It is an apt word for the beta-test-stage "Ur-functions" of Smith's fantastical Kindle, indicating that these are only the initial stages the product will take. But it also posits the digital Kindle, rather than the book, as the ur-medium, the origin of all literary culture.

Temporal boundaries are further destabilized, setting up the dramatic climax of the narrative, when Smith realizes that the experimental functions also allow him to access newspapers from the future. Even more troubling to him than the Ur in which the entire world is decimated by a 1962 nuclear holocaust is a local news story describing his estranged girlfriend's death in a traffic accident. The rest of the novella consists of Smith's work to avert that fate. He is successful, and the book ends with the couple's happy reunification. He also manages to escape the wrath of the Paradox Police, who come to punish him for altering his future, an act that violates the Paradox Laws. They overlook his transgression after they admit that he was mistakenly shipped a Kindle from an alternate Ur. In King's parable of the future of digital reading, the mythos of the Kindle is unsustainable. *UR*'s multiple potential futures, multiple potential universes, and multiple potential libraries cannot coexist. The Paradox Police take the Kindle, leaving Smith firmly returned to his own familiar, if limited, reality.

Smith's Kindle is uncanny, unsettling the foundations of reality and time. Either the device or the present moment must go. King has voiced his own sense that e-readers, as a category, possess a similar uncanniness: "There's a troubling lightness to the[ir] content. . . . A not-thereness. Even formidable works . . . feel somehow not quite real when read on a screen."[89] This screen essentialism ignores the extent to which digital data are both embodied and durable, but statements such as King's remain central to the imaginary of reading as it is shaped by different textual media. *UR* also foregrounds the uncanniness of e-readers by juxtaposing the fictional *UR*-Kindle with the real reader's actual Kindle. Nonfictional Kindles do not come in pink, nor do their menus contain Ur-functions. Nonetheless, King's descriptions of the device's menus and navigation are otherwise accurate, and discussions within the narrative of how to download texts and turn pages correctly describe the actions that the novella's reader has used to download and flip through the virtual pages of *UR*. As readers hold their own Kindles—as they read the words on their Kindle screens that Wesley Smith reads on his—they experience the uncanniness of the medial *mise en abyme* that I have shown to be a recurrent feature in the meta-medial novel.[90] This frisson between the fictionality of the represented medium and its grounding in the actual medium that stores it gestures simultaneously to the material experience of the Kindle and to King's perception of electronic textuality's "not-thereness."

An alternate take on the Kindle's "not-thereness" comes via *56 Broken Kindle Screens* (2012), an artist's book produced by the German artist Sebastian Schmieg in collaboration with the Italian artist and designer Silvio Lorusso. The work, which contains images of exactly what the title specifies, is available as a print-on-demand paperback or as an e-book. (The artists recommend that readers "jailbreak" their Kindle.) As Schmieg and Lorusso describe it, *56 Broken Kindle Screens* "takes advantage of the peculiar aesthetic of broken E Ink displays and serves as an examination into the reading device's materiality. As the screens break, they become collages composed of different pages, cover illustrations, and interface elements."[91] Most obviously, *56 Broken Kindle Screens* emphasizes the fragility of the Kindle. This platform might hold the future of reading, as Schmieg and Lorusso imply, but any individual Kindle may break at any time. Screens break commonly enough for the artists to have culled dozens of examples from Internet posts.

Viewed as a print book, *56 Broken Kindle Screens* juxtaposes the relative stability of the book's display with the fragility of the Kindle while also stressing the spatiality of codex reading. Many of the images pair dynamically across page spreads, generating meaningful interactions between the pictures that do not occur when the work is read as a series of sequential images on a Kindle. In one spread, the left page shows a woman's pale face close to the margin, the rest of the image obscured by intersecting vertical and horizontal white lines (26). On the adjacent page, a woman's pale hand is visible against a dark background, as if reaching out of window; the rest of the page is obscured by similar white lines (27). Taken together, the two seem to form the image of a single woman, emerging out of the recesses of the book. Codex-specific juxtapositions encourage the reader to seek such connections among the images, producing impressionistic narratives about their meaning. The actual Kindle hardware is relegated to the very back of the book: the only time we view anything beyond the boundaries of the Kindle screen is in Schmieg and Lorusso's unconventional reference list, which contains pictures of each of the previous images, framed by the broken Kindles that housed them, under which appear the URLs where the artists found the images online. The paperback book eclipses the Kindle reader, collapsing screen and page as display interfaces while demonstrating the particular relationships that the codex imparts on its contents.

234

Viewed on a Kindle, *56 Broken Kindle Screens* creates another uncanny medial *mise en abyme* as the images of the broken display fill the reader's own screen. Using E Ink to reproduce images of nonfunctioning E Ink displays, the digital version proleptically rehearses the possibility that the reader's Kindle will meet a similar fate. Schmieg and Lorusso also establish the Kindle's unreadability more subtly. Print culture is disproportionately fetishized in the images—*56 Broken Kindle Screens* includes pictures of famous writers such as Ralph Ellison and Virginia Woolf, a picture of a writing desk, and a closeup of type letterforms—but there is little actual text. Of the nearly five dozen images, only a handful display narrative text—that is, text from the contents of whatever e-book the reader might actually want to read on the Kindle. Instead, we see textual artifacts such as Kindle's operating instructions, a navigation menu, a low-battery warning, and, with appropriate irony, two advertisements for the Kindle. Thus, *56 Broken Kindle Screens* enacts what Roger Chartier has described as the manner in which electronic media collapse spatiality and defy a sense of volumetric scale. When one reads from a digital device, all reading matter occupies the same space, that of the screen: "In the digital world, all texts, whatever their genre, are produced or received through the same medium and in very similar forms."[92] Schmieg and Lorusso's work also drives home the degree to which interactions with surfaces continue to dominate perceptions of electronic media, both in how the images keep the focus on the screen as the surface of reading and in the compilers' selection of images that we might think of as digital "covers"—the Kindle home screen pictures rather than the text contained within any individual work. Like the Future Library and *UR*, the project emphasizes the complex intermingling of media cultures. The three also fortify a vision of a future in which the print book persists, whether as a concertedly retro but stable old medium (the Future Library), as a more substantial and less dangerous alternative to the digital (*UR*), or as a set of aesthetics whose material basis may outlive the digital platforms that adopt them (*56 Broken Kindle Screens*).

MODERNISM, LITERARY MEDIA HISTORY, AND THE TIME OF THE BOOK

I have spent much of this chapter assessing how the future is being imagined and constructed in the early decades of the twenty-first century, a time when the post-ness of the print book has been, for several decades, announced ad nauseum and mourned even as the book repeatedly survives. I want to end by turning back to the scene with which I opened this volume: D. H. Lawrence's contention that John Dos Passos's literary method makes the novel into a camera and a sound recorder so that "the book"—*Manhattan Transfer* and, more broadly, the print novel—"becomes what life is."[93]

As I have shown, modernist writers such as Dos Passos, Woolf, and Joyce laid the foundation for today's metamedial writing. They created a template for the twenty-first-century writers who build on their work of examining how the novel's archival project has dovetailed with the affordances of the print book while interrogating the claim that its unique recording propensities have grown obsolete with the development of newer information storage media. These modernist contributions to the theorization of information management and mediation represent one channel by which modernist writing retains its vital capacity to illuminate our thinking about the world. Many of its hallmarks continue to influence writers and scholars. Experimental authors from hypertext novelists to conceptual poets have invoked modernist writers as their literary forerunners.[94] Many of literary criticism's current practices—attention to form, interest in the writer's use of the medium, and the push beyond surface reading in favor of more symptomatic interpretation—are grounded in practices established in the previous century's first decades, even if the ideological grounds of literary critique have shifted dramatically in the intervening years. Modernist studies, too, has been revitalized since the turn of the twenty-first century with criticism that extends the field beyond the tight constraints that originally defined its objects of analysis.

Modernist criticism has expanded globally as well as vertically through time, encompassing non-European contexts and considering how later decades might be understood as modernist.[95] Scholars of modernism have challenged its dominant critical narratives, eschewing traditional formulations (such as the divide between elite high modernism and popular culture) in favor of examining "bad" or "weak" modernisms.[96] Studies of the period have also moved laterally into the print culture of the time. If such matter is not properly the subject of modernist studies per se, as Patrick Collier has argued, it nonetheless enriches the critical reception of modernity's media culture.[97]

Modernist studies has made a strong case for the relevance of the modernist period to the contemporary world. One consistent note in the many critical arguments that modernism prefigured some aspect of life in the twenty-first century is that modernity endured unprecedented social, political, and cultural upheavals. Even as mass culture and avant-garde writing offered novel forms of novelty, the era was bracketed by declines. Citing such factors as "a concern with the downturn of empire," "a registering of the eclipse of British and Anglo-Irish aristocracies," and "belief in a more general or abstract exhaustion of civilization," Douglas Mao argues that "we can notice that in the years around 1900, attributes of decadence in a world-historical sense clustered especially thickly around Great Britain."[98] The language of modernist writers spoke to the uncertainty of the future at a time when destruction was the guiding principle. It was an age of "breaking and falling, crashing and destruction," as Woolf famously wrote.[99] Because temporality, historicism, and futurity were concepts under revision during this period, modernist scholars have asserted that the work of modernism is essential to the understanding of late twentieth- and early twenty-first-century precarity, whether due to nuclear war, climate change, or other existential threats.[100]

For all of these reasons, literary modernism's entanglement with its media culture is compelling for contemporary thought about the interconnections and interactions between literature and media.

Discussions of digital media's impact are marked by a similar rhetoric of decline, transition, and rupture. The metamedial modernist novels I have analyzed contend that the book remains resonant because of, not despite, the existence of new scales of information that call into question not only the place of the book but also the place of the human in the knowledge economy. The theorization of modernism and modernity has benefited from an intensification of critical interest in modernism's media culture. Many monographs in this category— among them, Mark Goble's *Beautiful Circuits: Modernism and the Mediated Life,* David Trotter's *Literature in the First Media Age: Britain between the Wars,* and John Lurz's *The Death of the Book: Modernist Novels and the Time of Reading*—end by looking ahead to the digital moment, a move quickly becoming a critical trope. "Modernism as we know it," concludes Goble, "is not going to return. But the digital technologies that are the future also look a lot like history, and maybe there is still modernism enough to hunt us back to mediums that we have left behind."[101] Writing with an eye to their subjects' applicability to twenty-first-century media culture, these scholars indicate further lines of critical inquiry, junctures at which the assumptions and ideologies of the medial and literary forms circulating during the modernist period persist.

In structuring my own argument dialogically, moving back and forth between modernism and the contemporary moment, I have attempted to open more of these connections and to show where our thinking about digital mediation and the stakes of print books remains indebted to earlier ideas about the role of form in information culture and literary production. As scholars attend to aspects of modernism's media culture that have become newly visible due to their similarities with the twenty-first-century media episteme, we are better able to uncover the antecedents of digital culture as well as to theorize more comprehensively the ways in which digital media set out historically new conditions. Much research remains to be done to fully account for the print book's influence on the history of the novel, and I hope that the flourishing interest in and interpenetration of fields such as book history, media studies, and genre theory will enable this necessary work to thrive. I have focused on the early twentieth and twenty-first centuries because they are unusually energized in their confrontations of the book's status in a changing information ecology and in their literary responses to these changes. Future research should also look at different eras—the late 1800s, the 1960s, and other periods when the materiality of the book and its ability to harness the cultural impact of media and information has mattered to the formation of the novel.

The age of the novel has passed. So, too, the age of the book. The novel is still a privileged form in literary studies, but it no longer holds the lion's share of the public's consumption of narrative. The print book also endures, but it, too, has passed from the default medium for storing textual information to one among many options, increasingly outpaced by the sheer volume of information stored and communicated using digital technology. The most interesting analyses of media, however, have always understood that a media ecology is not a zero-sum environment. As Derrida writes in "The Book to Come":

> There is, there will therefore be, as always, the coexistence and structural survival of past models at the moment when genesis gives rise to new possibilities. . . . I'm in love with the book, in my own way and forever. . . . But I also love . . . the computer and the TV. And I like writing with a pen just as much, sometimes just as little, as writing with a typewriter. . . . A new economy is being put in place. It brings into coexistence, in a mobile way, a multiplicity of models, and of modes of archiving and accumulation. And that's what the history of the book has always been.[102]

Residual media matter: as holdovers from the past, participants in the present moment, and, in their persistence, agents of the future. Even those platforms most promising for digital reading—those, like the Kindle, that avoid the physical side-effects of screen reading and the distraction of wired environments—do not so much replace the book as coexist alongside it. Precisely because of their more marginal position, novels and books, and metamedial novels as they merge the two, can speak cogently to the conditions and consequences of media transition. As we reflect on our own media age, and as we look ahead to the future, we would do well to acknowledge the novel's archival project and to consider how the study of the novel and the book can help us understand our own drive to inhabit our present and ensure our future.

NOTES

Preface

1. Quoted in Frank Budgen, *James Joyce and the Making of "Ulysses"* (Bloomington: Indiana University Press, 1960), 67–68.
2. Filippo Marinetti, "The Founding and Manifesto of Futurism" (1909), in *Futurism: An Anthology,* ed. Lawrence Rainey, Christine Poggi, and Laura Wittman (New Haven, CT: Yale University Press, 2009), 52.
3. Walter Benjamin, "One-Way Street" (1928), in *One-Way Street and Other Writings,* trans. Edmund Jephcott and Kinglsey Shorter (London: NLB, 1979), 63.
4. Alexander Starre coined the term *metamedial*; see his *Metamedia: American Book Fictions and Literary Print Culture after Digitization* (Iowa City: University of Iowa Press, 2015).

Introduction: Information beyond the Book

1. D. H. Lawrence, "Review [of *Manhattan Transfer,* by John Dos Passos]" (1927), in *John Dos Passos,* ed. Barry Maine (New York: Routledge, 1988), 71–72, 71.
2. Ian Watt, *The Rise of the Novel: Studies in Defoe, Richardson, and Fielding* (Berkeley: University of California Press, 2001), 32–33.
3. E. M. Forster, "Anonymity: An Enquiry," *Calendar of Modern Letters* 2, no. 9 (1925): 148.
4. Lawrence, "Review," 71.
5. By "book" I mean print codex. While medium is the term commonly used to discuss books, strictly speaking a book is not a medium so much as a format, entailing medium (print or other textual inscription), support (paper), and platform (codex). I will have more to say about these distinctions over the course of this book.
6. I am drawing on Jay David Bolter and Richard A. Grusin's definition of transparent media in *Remediation: Understanding New Media* (Cambridge, MA: MIT Press, 2000). Bolter and Grusin define a transparent interface as "one that erases itself, so that the user is no longer aware of confronting a medium" (24).
7. Raymond Williams, "Dominant, Residual, and Emergent," in *Marxism and Literature* (Oxford: Oxford University Press, 1997), 122.
8. Bolter and Grusin, *Remediation,* 272. The subject of Bolter and Grusin's analysis is the history of visual representation in western culture, and they describe hypermediacy as "a style of visual representation," one that "multiplies the signs of mediation" by containing multiple media within the same visual field, as

in a computer desktop interface or a newspaper's "patchwork layout" (272, 32, 41). My use of their term does not entail this multimedia visual aesthetic—although several of the novels I analyze do adopt this visual style—but rather its effect: "In every manifestation, hypermediacy makes us aware of the medium" (34).

9. *Affordance* is a term from design theory; widely used in media studies, it has gained traction in literary studies via Caroline Levine's groundbreaking monograph *Forms: Whole, Rhythm, Hierarchy, Network* (Princeton, NJ: Princeton University Press, 2015). As she writes, the term is "used to describe the potential uses or actions latent in material and designs" (6). The language of affordance usefully moves away from strictly determinist accounts of media to discussion of conventions, defaults, and potentiality.

10. See, for example, N. Katherine Hayles, *Writing Machines* (Cambridge, MA: MIT Press, 2002).

11. Although such novels exist in languages other than English, I focus on Anglophone texts from the United Kingdom, Ireland, and the United States. While my analysis is attentive to the differences in the media and information cultures of each national context, part of the work of *Out of Print* is to establish the many factors they have had in common, creating a transnational framework for the study of information and media history as they pertain to the novel.

12. N. Katherine Hayles defines *media-specific analysis* as a mode that "attends both to the specificity of the form" of a text as it is embodied in a medium "and to citations and imitations of one medium in another" ("Print Is Flat, Code Is Deep," *Poetics Today* 25, no. 1 [2004]: 69). Media-specific analysis is concerned both with the particularity of individual media and with the positioning of these media within media ecologies.

13. Kate Marshall, *Corridor: Media Architectures in American Fiction* (Minneapolis: University of Minnesota Press, 2013), 16.

14. For example, J. Paul Hunter argues that "the fact that novels are conceived within the expectations and possibilities of the print medium is . . . crucial . . . because any discourse can more readily expand in a printed version, move in more directions, stay longer, go on less predictable tangents, and become a focus of attention for long periods of time without endangering the narrative center" (*Before Novels: The Cultural Contexts of Eighteenth-Century Fiction* [New York: Norton, 1990], 53).

15. Levine, *Forms,* 13.

16. Marco Codebò defines the *archival novel* as "a fictional genre where the narrative stores records, bureaucratic writing informs language, and the archive functions as a semiotic frame" for the narrative (*Narrating from the Archive: Novels, Records, and Bureaucrats in the Modern Age* [Madison, NJ: Fairleigh Dickinson University Press, 2010], 13). On the encyclopedic, see Edward Mendelson, "Encyclopedic Narrative, from Dante to Pynchon," *MLN* 91, no. 6 (1976): 1267–75.

17. Arthur Bahr, *Fragments and Assemblages: Forming Compilations of Medieval London* (Chicago: University of Chicago Press, 2012), 218.

18. E. M. Forster, *Aspects of the Novel* (London: Harcourt Brace, 1927), 25; Jerome McGann, *Radiant Textuality: Literature after the World Wide Web* (New York: Palgrave, 2001), 178.

19. Watt, *Rise of the Novel*, 32–33.

20. Hunter, *Before Novels*, 32.

21. In her work on literature and the theory of mind, Lisa Zunshine describes the novel "as a sustained representation of numerous interacting minds" and thus the best genre for studying how fictional minds accord with real cognitive processes (*Why We Read Fiction: Theory of Mind and the Novel* [Columbus: Ohio State University Press, 2006], 10).

22. See Friedrich A. Kittler, *Gramophone, Film, Typewriter*, trans. Geoffrey Winthrop-Young and Michael Wutz (Stanford, CA: Stanford University Press, 1999).

23. Lisa Gitelman and Virginia Jackson, "Introduction," in *"Raw Data" Is an Oxymoron*, ed. Lisa Gitelman (Cambridge, MA: MIT Press, 2013), 1. Definitions of data vary greatly by field. As Christine L. Borgman puts it, "the only agreement on definitions [of data] is that no single definition will suffice. . . . The value of data varies widely over place, time, and context. . . . Data have no value or meaning in isolation" (*Big Data, Little Data, No Data: Scholarship in the Networked World* [Cambridge, MA: MIT Press, 2015], 4). For a history of the word *data*, see Daniel Rosenberg's "Data before the Fact" (in Gitelman, *"Raw Data" Is an Oxymoron*, 15–40). For a thorough overview of different definitions of data, see chapter 2 of Borgman's *Big Data, Little Data, No Data*.

24. Borgman, *Big Data, Little Data, No Data*, 28.

25. "Knowledge," *Oxford English Dictionary Online*.

26. Ann M. Blair, *Too Much to Know: Managing Scholarly Information before the Modern Age* (New Haven, CT: Yale University Press, 2010), 2.

27. John Guillory, "The Memo and Modernity," *Critical Inquiry* 31, no. 1 (2004): 110; "information," *Oxford English Dictionary Online* (this is the sense used by Claude Shannon in his development of information theory); Gregory Bateson, *Steps to an Ecology of Mind* (Chicago: University of Chicago Press, 1972), 315.

28. For instance, John Seely Brown and Paul Duguid write that "knowledge usually entails a knower. That is, where people treat information as independent and more-or-less self-sufficient, they seem more inclined to associate knowledge with someone" (*The Social Life of Information* [Cambridge, MA: Harvard Business Review Press, 2000], 119).

29. Forster, "Anonymity," 148.

30. Watt, *Rise of the Novel*, 13; Erich Auerbach, *Mimesis: The Representation of Reality in Western Literature*, trans. Willard R. Trask (Princeton, NJ: Princeton University Press, 1953), 49; Mikhail Mikhaĭlovich Bakhtin, *The Dialogic Imagination* (1981),

ed. Michael Holquist, trans. Caryl Emerson and Michael Holquist (Austin: University of Texas Press, 2004), 39; Virginia Woolf, "A Room of One's Own" (1929), in *Selected Works of Virginia Woolf* (London: Wordsworth Editions, 2012), 611; Terry Eagleton, *The English Novel: An Introduction* (Oxford: Blackwell, 2005), 5–6; Michael McKeon, "Genre Theory," in *Theory of the Novel: A Historical Approach,* ed. Michael McKeon (Baltimore: Johns Hopkins University Press, 2000), 4.

31. For a history of how the novel arose among genres such as the romance, the ballad, and the news pamphlet, which also complicated the fact-fiction distinction, see Lennard J. Davis, *Factual Fictions: The Origins of the English Novel* (Philadelphia: University of Pennsylvania Press, 1996).

32. The bookkeeping log is from B. S. Johnson's novel *Christie Malry's Own Double-Entry.* I discuss the kitchen-shelf items and other lists in *Ulysses* in chapter 1.

33. Jessica Pressman, "The Aesthetic of Bookishness in Twenty-First-Century Literature," *Michigan Quarterly Review* 48, no. 4 (2009): 465–82.

34. Eli Pariser, *The Filter Bubble: What the Internet Is Hiding from You* (New York: Penguin, 2011), 11.

35. On ubiquitous computing and recent interface ideologies, see Lori Emerson, *Reading Writing Interfaces: From the Digital to the Bookbound* (Minneapolis: University of Minnesota Press, 2014), especially chapter 1.

36. The origins of the phrase are obscure. Invoking something like its current connotations, the phrase "big data" began to be used in conversations in the tech community in the mid-1990s, and it appeared in the title of a paper at least as early as 1997. It had begun to appear in published academic works by 2003; in 2008, *Wired* magazine popularized the term. In the second decade of the twenty-first century, the term's popularity and usage increased exponentially. By 2013, Big Data had an entry in the *Oxford English Dictionary*.

37. Borgman, *Big Data, Little Data, No Data,* 4.

38. danah boyd and Kate Crawford, "Critical Questions for Big Data: Provocations for a Cultural, Technological, and Scholarly Phenomenon," *Information, Communication, and Society* 15, no. 5 (2012), 663. I follow boyd's preferred convention for writing her name.

39. "Big data," *Oxford English Dictionary Online,* emphasis added.

40. Whereas deep attention "is characterized by concentrating on a single object for long periods," hyper attention "is characterized by switching focus rapidly among different tasks, preferring multiple information streams, seeking a high level of stimulation, and having a low tolerance for boredom" (N. Katherine Hayles, "Hyper and Deep Attention: The Generational Shift in Cognitive Modes," *Profession* [2007]: 187).

41. Nicholas Carr, *The Shallows: What the Internet Is Doing to Our Brains* (London: Norton, 2011), 103; Sven Birkerts, *The Gutenberg Elegies: The Fate of Reading in an*

Electronic Age, rev. ed. (New York: Faber and Faber, 2006), 18; Hayles, "Hyper and Deep Attention," 187–88.

42. *Pace* Carr and Birkerts, mobile media contribute to the dissemination of literary texts, and dedicated e-book readers show that digital platforms, too, may be designed to encourage deep attention. On the rise of deep reading, see David Dowling, "Escaping the Shallows: Deep Reading's Revival in the Digital Age," *Digital Humanities Quarterly* 8, no. 2 (2014): n.p.

43. Forster, *Aspects of the Novel,* 6; Naomi S. Baron, *Words Onscreen: The Fate of Reading in a Digital World* (New York: Oxford University Press, 2015), 48.

44. Walter Benjamin, "The Storyteller" (1936), in *Illuminations: Essays and Reflections,* ed. Hannah Arendt, trans. Harry Zohn (New York: Schocken, 1968), 87.

45. Lev Manovich, *The Language of New Media* (Cambridge, MA: MIT Press, 2001), 225. *Natural symbionts* is N. Katherine Hayles's term; she contends that, "because database can construct relational juxtapositions but is helpless to interpret or explain them, it needs narrative to make its results meaningful. Narrative, for its part, needs database in the computationally intensive culture of the new millennium to enhance its cultural authority and test the generality of its insights" ("Narrative and Database: Natural Symbionts," *PMLA* 122, no. 5 (2007): 1603).

46. Michael Wutz, *Enduring Words: Literary Narrative in a Changing Media Ecology* (Tuscaloosa: University of Alabama Press, 2009), 24. For Wutz, this difference is the reason for the novel's continued relevance rather than a factor hastening its obsolescence.

47. See Hayles, *Writing Machines;* Alison Gibbons, *Multimodality, Cognition, and Experimental Literature* (New York: Routledge, 2012); and Alexander Starre, *Metamedia: American Book Fictions and Literary Print Culture after Digitization* (Iowa City: University of Iowa Press, 2015).

48. I refer to Salvador Plascencia's *The People of Paper* (San Francisco: McSweeney's, 2005), Steven Hall's *The Raw Shark Texts* (Edinburgh: Canongate, 2007), and J. J. Abrams's and Doug Dorst's *S.* (New York: Mulholland, 2013).

49. Leah Price, "Introduction: Reading Matter," *PMLA* 121, no. 1 (2006): 12; Kiene Brillenburg Wurth, "Book Presence: An Introductory Exploration," in *Book Presence in a Digital Age,* ed. Kiene Brillenburg Wurth, Kári Driscoll, and Jessica Pressman (New York: Bloomsbury Academic, 2018), 8.

50. Alexander Starre, "'Little Heavy Papery Beautiful Things': McSweeney's, Metamediality, and the Rejuvenation of the Book in the US," *Writing Technologies,* vol. 3 (2010): 32. See also his *Metamedia.*

51. My thanks to Kelley Kreitz for suggesting the idea of novels as alternative media.

52. "Interface," *Oxford English Dictionary Online.* For a typology of computational ·interfaces, see Florian Cramer and Matthew Fuller, "Interface," in *Software*

Studies: A Lexicon, ed. Sean Cubitt and Roger F. Malina (Cambridge, MA: MIT Press, 2008), 149–52.

53. Bonnie Mak, for instance, describes a page as "an interface, standing at the centre of the complicated dynamic of intention and reception" (How the Page Matters [Toronto: University of Toronto Press, 2011], 21). Compare McGann: "A page of printed or scripted text should . . . be understood as a certain kind of graphic interface" (Radiant Textuality, 199).

54. Emerson, Reading Writing Interfaces, 6. My work joins Emerson's in thinking about the way in which literature can expose how "the glossy surface of the interface . . . alienates the user from having access to the underlying workings of the device" (xi). Emerson focuses on works that demystify the involvement of interfaces in writing.

55. For instance, Daniel Punday's Writing at the Limit: The Novel in the New Media Ecology focuses on contemporary novels that contain "references to other media (rather than the use of those other media)," purposefully excluding metamedial works ([Lincoln: University of Nebraska Press, 2012], 26). While his monograph provides a comprehensive study of the ways in which "written narrative" differs from other media such as film, painting, and music, there is only passing consideration of how either print or the book influences written narrative (37). In a similar vein, Tony E. Jackson's The Technology of the Novel: Writing and Narrative in British Fiction describes the novel as "a literary form that has been primarily associated with the technology of print" but does not distinguish print from writing in the subsequent analysis ([Baltimore: Johns Hopkins University Press, 2009], 2).

56. These terms come from Starre, Metamedia, and Gibbons, Multimodality.

57. Marshall, Corridor; John Lurz, The Death of the Book: Modernist Novels and the Time of Reading (New York: Fordham University Press, 2016); Wutz, Enduring Words.

58. Jonathan Sterne, MP3: The Meaning of a Format (Durham, NC: Duke University Press, 2012), 7.

59. Amaranth Borsuk, The Book (Cambridge, MA: MIT Press, 2018), 111.

60. Johanna Drucker, The Visible Word: Experimental Typography and Modern Art, 1909–1923 (Chicago: University of Chicago Press, 1994), 95.

61. For this reason, Hayles describes "print as a particular form of output for electronic text" (My Mother Was a Computer: Digital Subjects and Literary Texts [Chicago: University of Chicago Press, 2005], 117).

62. Robert Coover, "The End of Books," New York Times, June 21, 1992.

63. For these examples, see Timothy C. Campbell, Wireless Writing in the Age of Marconi (Minneapolis: University of Minnesota Press, 2006); Mark Goble, Beautiful Circuits: Modernism and the Mediated Life (New York: Columbia University Press, 2000); Marshall, Corridor; and David Trotter, Literature in the

First Media Age: Britain between the Wars (Cambridge, MA: Harvard University Press, 2013).

64. Friedrich A. Kittler, *Discourse Networks, 1800/1900,* trans. Michael Metteer with Chris Cullens (Stanford, CA: Stanford University Press, 1992).

65. The "new" modernist studies are by now well established, dating roughly to the founding of the Modernist Studies Association (1998) and the journal *Modernism/modernity* (1994). See Douglas Mao and Rebecca L. Walkowitz, "The New Modernist Studies," *PMLA* 123, no. 3 (2008): 737–48.

66. See Manovich, *The Language of New Media;* Alan Liu, "Transcendental Data: Toward a Cultural History and Aesthetics of the New Encoded Discourse," in *Local Transcendence: Essays on Postmodern Historicism and the Database* (Chicago: University of Chicago Press, 2008), 210; Paul Stephens, *The Poetics of Information Overload: From Gertrude Stein to Conceptual Writing* (Minneapolis: University of Minnesota Press, 2015); and Jessica Pressman, *Digital Modernism: Making It New in New Media* (Oxford: Oxford University Press, 2014).

67. Alexis Weedon, *Victorian Publishing: The Economics of Book Production for a Mass Market, 1836–1916* (Aldershot, UK: Ashgate, 2003), 57. Janice Radway, "Research Universities, Periodical Publication, and the Circulation of Professional Expertise: On the Significance of Middlebrow Authority," *Critical Inquiry* 31, no. 1 (2004): 216.

68. "Literary Gossip," *Athenaeum,* December 14, 1912, 731; Clifford Smyth, "The Fiction Famine," *Literary Digest International Book Review* 1 (October 1923): 22.

69. Walter Benjamin, "One-Way Street" (1928), in *One-Way Street and Other Writings,* trans. Edmund Jephcott and Kinglsey Shorter (London: NLB, 1979), 61–62.

70. See Hayles, *My Mother Was a Computer,* 31–32, 117–18.

71. Marshall, *Corridor,* 28.

72. William Carlos Williams, *Imaginations* (New York: New Directions, 1970), 85.

73. Borsuk, *The Book,* 112, 113; the quotations describe artists' books.

Chapter 1: Information Shock

1. Virginia Woolf, *Mrs. Dalloway* (1925; London: Harcourt, 1981), 20–21.

2. Skywriting was a new phenomenon when Woolf was writing *Mrs. Dalloway,* and it was closely associated with aerial warfare. As Paul K. Saint-Amour explains, "[it] was developed during the First World War by the British flying ace J. C. Savage, who patented the technology; his company, Savage Skywriting, was the first to deploy it commercially, writing 'Castrol,' 'Daily Mail,' and 'Persil' over Derby in May 1922" (*Tense Future: Modernism, Total War, Encyclopedic Form* [Oxford: Oxford University Press, 2015], 114n). In *Mrs. Dalloway,* the shell-shocked veteran Septimus Smith also views the skywriting, making explicit the plane's connection to the First World War.

3. "Too Much to Read" [advertisement], *Weekly Review,* January 31, 1920, iii.

4. Henry Seidel Canby, "Current Literature and the Colleges," *Harper's Monthly Magazine,* 131 (June–November 1915): 234.

5. Many histories of information science locate the field's beginnings in the aftermath of the Second World War. Vannevar Bush's 1945 essay "As We May Think" is widely cited as bringing to light the need for systematic information organization as a postwar phenomenon in the United States. Similar changes took place in the United Kingdom. MI5, the British security service, was so inundated with information that, as one account puts it, "at the start of the Second World War MI5's information management system virtually collapsed due to a tidal wave of data flooding the system" (Alistair Black and Dave Muddiman, "The Information Society before the Computer," in *The Early Information Society: Information Management in Britain before the Computer,* ed. Alistair Black, Dave Muddiman, and Helen Plant [Aldershot, UK: Ashgate, 2007], 29).

6. P. C. Judge, "Adrift on an Ocean of Print: A Reader's Perplexity," *Irish Independent,* May 17, 1923, 6.

7. Ann M. Blair, *Too Much to Know: Managing Scholarly Information before the Modern Age* (New Haven, CT: Yale University Press, 2010), 55, 15.

8. "Too Many Books," *Evening Herald,* January 19, 1927, 5.

9. In the United States, rates of illiteracy fell from 20 percent in 1870 to less than 5 percent in 1930 across the total population, although a significant racial gap remained (National Center for Education Statistics, *120 Years of American Education: A Statistical Portrait,* ed. Thomas D. Snyder [Washington, DC: U.S. Department of Education, 1993], 21). By the turn of the twentieth century, nearly 100 percent of the population of England, Wales, and Scotland could read (Joseph McAleer, *Popular Reading and Publishing in Britain, 1914–1950* [Oxford: Clarendon, 1992], 14). In Ireland, the impacts of poverty, the death toll from the Great Famine emigration, and the language transition from Irish to English meant that literacy rates grew less dramatically. However, 86 percent of the population in Ireland could read in English in 1901, as compared with only 53 percent fifty years earlier (Clare Hutton, "Publishing the Irish Cultural Revival, 1891–1922," in *The Oxford History of the Irish Book*, vol. 5, *The Irish Book in English, 1891–1922,* ed. Clare Hutton and Patrick Walsh [Oxford: Oxford University Press, 2011], 19–21); Hutton notes that "almost all monoglot speakers of Irish were illiterate" (20).

10. On the history of Irish publishing during this period, see Hutton, "Publishing the Irish Cultural Revival." On Ireland's reputation as a country in which "the existence of a substantial book-reading and book-buying public" "was largely missing," see Tony Farmar, "An Eye to Business: Financial and Market Factors, 1895–1995," in Hutton and Walsh, *The Oxford History of the Irish Book,* 5:211.

11. As Melba Cuddy-Keane notes, whereas such concerns had, in the nineteenth century, been associated with the rise of "*functional* literacy," by the modernist period they resulted in the more complex problem of "*cultural* literacy" (*Virginia Woolf, the Intellectual, and the Public Sphere* [Cambridge: Cambridge University Press, 2003], 60).

12. Quoted in Walter Benjamin, "The Work of Art in the Age of Mechanical Reproduction," in *Illuminations: Essays and Reflections,* ed. Hannah Arendt, trans. Harry Zohn (New York: Schocken, 1968), 248.

13. Leonard Woolf and Virginia Woolf, "Are Too Many Books Written and Published?" (1927), intro. Melba Cuddy-Keane, *PMLA* 121, no. 1 (2006): 241.

14. Ibid., 240.

15. "The Best Advice on Books" [advertisement], *Bermondsey Book* 6, no. 1 (1928–29): 119.

16. Patrick Collier, *Modern Print Artefacts: Textual Materiality and Literary Value in British Print Culture, 1890–1930s* (Edinburgh: Edinburgh University Press, 2018), 97.

17. Franklin T. Baker, "A Bibliography of Children's Reading," *Teacher's College Record* 9, no. 1 (1908): 7.

18. James Joyce, *Ulysses* (1922), ed. Hans Walter Gabler (New York: Vintage, 1986), 3.136–37. I follow the convention of citing *Ulysses* by episode and line number. Further page references appear in the text as parenthetical citations.

19. "Too Much to Read," iii; Canby, "Current Literature," 234.

20. Friedrich A. Kittler, *Gramophone, Film, Typewriter,* trans. Geoffrey Winthrop-Young and Michael Wutz (Stanford, CA: Stanford University Press, 1999).

21. Ann Ardis and Patrick Collier, "Introduction," in *Transatlantic Print Culture, 1880–1940: Emerging Media, Emerging Modernisms,* ed. Ann Ardis and Patrick Collier (Basingstoke, UK: Palgrave Macmillan, 2008), 1.

22. Janice A. Radway, "Research Universities, Periodical Publication, and the Circulation of Professional Expertise: On the Significance of Middlebrow Authority," *Critical Inquiry* 31, no. 1 (2004): 221.

23. "Too Much to Read," iii.

24. Alistair Black, "'A Valuable Handbook of Information': The Staff Magazine in the First Half of the Twentieth Century as a Means of Information Management," in *Information History in the Modern World: Histories of the Information Age,* ed. Toni Weller (Basingstoke, UK: Palgrave Macmillan, 2011), 149.

25. Ford Madox Ford [as Ford Madox Hueffer], *The Critical Attitude* (London: Duckworth, 1911), 125.

26. Carl F. Kaestle and Janice A. Radway, "A Framework for the History of Publishing and Reading in the United States, 1880–1940," in *A History of the Book in America,* vol. 4, *Print in Motion: The Expansion of Publishing and Reading in the*

United States, 1880–1940, ed. Carl F. Kaestle and Janice A. Radway (Chapel Hill: University of North Carolina Press, 2009), 12, 11.

27. James O'Toole and Sarah Smyth, *NEWSPLAN: Report of the NEWSPLAN Project in Ireland,* 2nd ed. (London and Dublin: British Library and National Library of Ireland, 1998), xviii.

28. Simon Eliot, "The Reading Experience Database; or, What Are We to Do about the History of Reading?," Open University, RED Project, http://www.open.ac.uk.

29. Walter Benjamin, "On Some Motifs in Baudelaire," in *Illuminations,* 175.

30. E. M. Forster, "Anonymity: An Enquiry," *Calendar of Modern Letters* 2, no. 9 (1925): 147.

31. Paul Stephens, *The Poetics of Information Overload: From Gertrude Stein to Conceptual Writing* (Minneapolis: University of Minnesota Press, 2015), 185.

32. Eliot, "Reading Experience Database."

33. Ronald E. Day, *Indexing It All: The Subject in the Age of Documentation, Information, and Data* (Cambridge, MA: MIT Press, 2014), 26.

34. Virginia Woolf, *Jacob's Room* (1922; Brooklyn: Melville House, 2011), 76.

35. Filippo Tommaso Marinetti, *Tribute to the Italian Guido Guidi Who, in an Italian Aircraft, Beat the World Height Record (7,950 M),* 1916. Ink, watercolor, and collage on paper. Private collection.

36. John Dos Passos, *Manhattan Transfer* (1925; Boston: Mariner, 2000), 41.

37. Ibid., 42.

38. Patrick A. McCarthy, "Reading in *Ulysses,*" in *Joycean Occasions: Essays from the Milwaukee James Joyce Conference, Volume 1987,* ed. Janet Egleson Dunelavy, Melvin J. Friedman, and Michael Patrick Gillespie (Cranbury, NJ: Associated University Presses, 1991), 18.

39. Joe Benton, "Advertising the School," *Printers' Ink,* March 4, 1908, 16.

40. On the potential benefits of hyper attention, see N. Katherine Hayles, "Hyper and Deep Attention: The Generational Divide in Cognitive Modes," *Profession* (2007): 187–99.

41. Alistair Black, Simon Pepper, and Kaye Bagshaw, *Books, Buildings, and Social Engineering: Early Public Libraries in Britain from Past to Present* (Farnham, UK: Ashgate, 2009), 32.

42. Wayne A. Wiegand, *Part of Our Lives: A People's History of the American Public Library* (New York: Oxford University Press, 2015), 104.

43. Ben Levitas, "Reading and the Irish Revival, 1891–1922," in Hutton and Walsh, *The Oxford History of the Irish Book,* 5:57–58.

44. Quoted in Alistair Black, "Networking Knowledge before the Information Society: The Manchester Central Library (1934) and the Metaphysical-Professional Philosophy of L. S. Jast," in *European Modernism and the Information*

Society: Informing the Present, Understanding the Past, ed. W. Boyd Rayward (Aldershot, UK: Ashgate, 2008), 169–70.

45. See Sianne Ngai, Ugly Feelings (Cambridge, MA: Harvard University Press, 2005), esp. 286–88.

46. Quoted in Black and Muddiman, "The Information Society," 16.

47. JoAnne Yates, Control through Communication: The Rise of System in American Management (Baltimore: Johns Hopkins University Press, 1989), xv.

48. E. R. Judders, Indexing and Filing: A Manual of Standard Practice (New York: Ronald Press Company, 1918), iii.

49. Cornelia Vismann, Files: Law and Media Technology (2000), trans. Geoffrey Winthrop-Young (Stanford, CA: Stanford University Press: 2008), 126, 125.

50. "Vertex Vertical-Expanding File Pockets" [advertisement], Filing and Office Management 6, no. 1 (1921): 2.

51. At the time of the Mundaneum's debut, Otlet and La Fontaine called the project the Palais Mundial—the "World Palace" or "World Center."

52. Otlet and La Fontaine viewed the Mundaneum as contributing to a new spirit of internationalism. Otlet would come to see it as facilitating the international cooperation that he hoped would provide the necessary foundation for world peace (a desire made intensely personal when one of his sons was killed in the First World War). Otlet and La Fontaine founded the Union of International Associations in 1907, out of which the League of Nations and the International Institute of Intellectual Cooperation (later combined with UNESCO) emerged. La Fontaine received the Nobel Peace Prize in 1913 for his other work in building international peace coalitions.

53. Union of International Associations, "The Union of International Associations: A World Centre" (1914), in International Organisation and Dissemination of Knowledge: Selected Essays of Paul Otlet, trans. and ed. W. Boyd Rayward (Amsterdam: Elsevier, 1990), 116.

54. Quoted in Barry James, "World of Learning and a Virtual Library," New York Times, June 27, 1998.

55. Following standard practice, I adopt the abbreviation RBU to reflect the original French title: Répertoire Bibliographique Universel.

56. W. Boyd Rayward, "Introduction," in International Organisation and Dissemination of Knowledge, 3. Today the IIB operates as the International Federation for Information and Documentation.

57. Alex Wright, "The Web That Time Forgot," New York Times, June 17, 2008; Kevin Kelley, quoted in ibid.

58. H. G. Wells outlined his concept for a permanent world encyclopedia in the 1930s, planning to combine "a world synthesis of bibliography and documentation with the indexed archives of the world" ("The Idea of a Permanent World

Encyclopedia" [1937], in *World Brain* [Freeport, NY: Books for Libraries Press, 1971], 85). Like the founders of the Mundaneum, he believed that information proliferation could be managed systematically: "Few people as yet . . . know how manageable well-ordered facts can be made, however multitudinous . . . once they have been put in place in a well-ordered scheme of reference and reproduction" (85–86).

59. Paul Otlet, "The Science of Bibliography and Documentation," in Rayward, *International Organisation and Dissemination,* 79.

60. From this point on, I refer primarily to Otlet, who produced the bulk of the writing about the Mundaneum and its bibliographic systems.

61. W. Boyd Rayward, "The Origins of Information Science and the International Institute of Bibliography/International Federation for Information and Documentation (FID)," *Journal of the American Society for Information Science* 48, no. 4 (1997): 295.

62. John Guillory, "The Memo and Modernity," *Critical Inquiry* 31, no. (2004): 126.

63. Paul Otlet, "Something about Bibliography," in Rayward, *International Organisation and Dissemination of Knowledge,* 17.

64. Quoted in W. Boyd Rayward, "Visions of Xanadu: Paul Otlet (1868–1944) and Hypertext," *Journal of the American Society for Information Science* 45, no. 4 (1994): 244, 240.

65. See Paul Otlet, "Manuel de la Documentation Administrative," *Publication de l'I.I.B* 137 (1923): 51.

66. Charles Greifenstein and Francis Palmer, *Filing as a Profession for Women* (Boston: Library Bureau, 1919), 28.

67. "Unique Plan for Reference Book; By Novel Binding Device Nelson's Encyclopedia Solves Problem of Perpetual Freshness," *New York Times,* January 4, 1908.

68. Otlet, "Something about Bibliography," 18.

69. Isabelle Rieusset-Lemarié, "P. Otlet's Mundaneum and the International Perspective in the History of Documentation and Information Science," in *Historical Studies in Information Science,* ed. Trudi Bellardo Hahn and Michael Keeble Buckland (Medford, NJ: Information Today, 1998), 38. Otlet also pictured the Mundaneum as a sphere, symbolizing its comprehensiveness and totality. This symbolism endures in the Mundaneum Museum, where a giant globe dominates the exhibition hall.

70. M. F. Donker Duyvis, "Third Report of the International Committee of Classification Decimal," *Publication de l'I.I.B.* no. 159 (1929): 9.

71. Yates, *Control through Communication,* 15.

72. Quoted in Robert V. Williams, "The Documentation and Special Libraries Movement in the United States, 1910–1960," *Journal of the American Society for Information Science* 48, no. 9 (1997): 173.

73. Day, *Indexing It All,* 41.

74. Otlet also imagined information interfaces visually similar to the computer screens of today: he proposed that future users of the Mundaneum would be able to access its information remotely, viewing documents on screens in their homes.

75. Alan Liu, "Transcendental Data: Toward a Cultural History and Aesthetics of the New Encoded Discourse," in *Local Transcendence: Essays on Postmodern Historicism and the Database* (Chicago: University of Chicago Press, 2008), 224.

76. "Special Libraries," *Leinster Express,* September 29, 1926.

77. Greifenstein and Palmer, *Filing as a Profession for Women,* 41.

78. Alex Wright, *Glut: Mastering Information through the Ages* (Ithaca: Cornell University Press, 2007), 189.

79. The Mundaneum's website quotes *Le Monde* describing the Mundaneum as "Google de papier."

80. Richard Kain, *Fabulous Voyager* (Chicago: University of Chicago Press, 1947), 37; Jacques Derrida, "*Ulysses,* Gramophone: Hear Say Yes in Joyce," in *Acts of Literature,* ed. Derek Attridge (New York: Routledge, 1992), 281. The claim that the task of documenting the city of Dublin might fall to literature rather than official institutions seems less hyperbolic in light of the fact that much of Dublin—including the city's archives—were, as Joyce was writing *Ulysses,* destroyed by the violence of the 1916 Easter Rising, the Irish War for Independence, and the Irish Civil War. In 1921, for example, the Irish Republican Army burned down Dublin's Custom House, destroying a significant collection of civic records: "All the most important documents of the Government relating to Ireland, together with papers and records, the value of which cannot be estimated, [were] destroyed" ("Sinn Feiners Burn Dublin Custom House," *New York Times,* May 29, 1921). On how *Ulysses* registers this violence, see Enda Duffy, *The Subaltern "Ulysses"* (Minneapolis: University of Minnesota Press), chap. 3; and Saint-Amour's *Tense Future,* chap. 5

81. James Joyce, *Finnegans Wake* (1939; New York: Penguin, 1999), 179.26–27.

82. Franco Moretti, *Modern Epic: The World-System from Goethe to García Márquez* (1994), trans. Quintin Hoare (London: Verso, 1996), 48.

83. See, for instance, Sara Danius's characterization of modernism as marked by ongoing "attempt[s] to turn the novel into an encyclopedic, totalizing, all-inclusive work of art" (*The Senses of Modernism: Technology, Perception, and Aesthetics* [Ithaca, NY: Cornell University Press, 2002], 4). Saint-Amour argues that Moretti's account of the modern epic overlooks the extent to which encyclopedism thrived during the modernist period. He demonstrates that, while encyclopedic novels such as *Ulysses* and Ford Madox Ford's *Parade's End* tetralogy share common features with the epic (namely, "massive scale," "radical inclusivity," and "ambitions to paint a comprehensive picture of national life"),

they resist the epic's totalizing unity (*Tense Future,* 185). Saint-Amour describes modernist encyclopedic form as an interwar phenomenon: he reads the encyclopedism of novels such as *Ulysses* as a challenge to the totalizing logic of the epic, a logic associated with the new possibility of total war. My reading of *Ulysses* accounts for its rejection of totalizing order from a different perspective: as a rejection of the logic of information mediation.

84. Stephens, *Poetics,* 54, 47. Stephens's argument that Stein rebelled against bureaucratic regimes of information organization through the use of literary forms that created states of distraction for readers is in keeping with this chapter's description of information shock as a condition to which modernist writers responded and a textual strategy they adopted. I am specifically concerned with how modernist writers implicated text printed in the pages of books in this formal experimentation as well as with how they understood this literary strategy within the context of the novel's archival impulse.

85. Gertrude Stein, *Tender Buttons: Objects, Food, Rooms* (1914; New York: Haskell House, 1970), 7, emphasis added.

86. In one of modernism's best-known examples of serialization, sections of *Ulysses* were published in the *Little Review* and the *Egoist* between 1918 and 1920. As Ed Mulhall points out, although this is the format in which Joyce's contemporaries first read *Ulysses,* the serialization represents a much earlier stage of the work, "almost a first draft of the novel" ("Ulysses' Journey: The First Sightings of James Joyce's Masterpiece," *Century Ireland* [March 2018]: 14). See also Clare Hutton, *Serial Encounters: Ulysses and the Little Review* (Oxford: Oxford University Press, 2019), esp. 85–86.

87. John Lurz, *The Death of the Book: Modernist Novels and the Time of Reading* (New York: Fordham University Press, 2016), 67.

88. T. S. Eliot, "*Ulysses,* Order, and Myth" (1923), in *Modernism: An Anthology,* ed. Lawrence Rainey (Malden: Blackwell, 2005), 167.

89. As Guillory notes, "informational writing . . . makes formatting as visible as possible"; with the advent of standardized forms in the late nineteenth century, there was a "shift from continuous prose to a graphically organized page" ("The Memo and Modernity," 128, 126).

90. Joyce's attention to books and fine printing traditions is not confined to "Ithaca." In "Circe," *Ulysses* mocks W. B. Yeats's sisters Elizabeth and Lily Yeats, who ran the Cuala Press, describing a book "to be printed and bound at the Druiddrum press by two designing females" with "calf covers of pissedon green" (14.1454–56).

91. M'Coy observed Bloom buy "a book from an old one in Liffey street for two bob" with "fine plates in it worth double the money"; the book is apparently the astronomy handbook mentioned in "Ithaca" (10.527–29). It is unclear whether

Bloom actually is a canny book buyer or whether this is M'Coy's (perhaps anti-Semitic) characterization of Bloom as materialistic.

92. Lawrence S. Rainey, *Institutions of Modernism: Literary Elites and Public Culture* (New Haven, CT: Yale University Press, 1998), 64.

93. According to Luca Crispi, Joyce recorded "tens of thousands of lexical notes for *Ulysses*" ("The Notescape of *Ulysses*," in *New Quotatoes: Joycean Exogenesis in the Digital Age,* ed. Ronan Crowley and Dirk Van Hulle [Leiden: Brill Rodopi, 2016], 78).

94. James Buzard, *Disorienting Fiction: The Autoethnographic Work of Nineteenth-Century British Novels* (Princeton, NJ: Princeton University Press, 2005), 25. Joyce would later refer to *Ulysses* as his "Blue Book of Eccles," aligning the novel with the British Parliamentary Papers popularly known as Blue Books (*Finnegans Wake*, 179.27).

95. Shane Leslie [as Domini Canis], review of *Ulysses, Dublin Review* 171 (September 1922): 117.

96. Caroline Levine, *Forms: Whole, Rhythm, Hierarchy, Network* (Princeton, NJ: Princeton University Press, 2015), 31.

97. Fritz Senn, "'Ithaca': Portrait of the Chapter as a Long List," in *Joyce's "Ithaca,"* ed. Andrew Gibson (Amsterdam: Rodopi, 1996), 52.

98. On "Ithaca" as a catalogue, see ibid. Andrew Gibson reads "Ithaca" in relation to the discourse of British science, including textbooks ("'An Aberration of the Light of Reason': Science and Cultural Politics in 'Ithaca,'" in Gibson, *Joyce's "Ithaca,"* 133–74). Jessica Pressman and Patrick Parrinder describe *Ulysses* as akin to "database" and "data-bank," respectively (Jessica Pressman, *Digital Modernism: Making It New in New Media* [Oxford: Oxford University Press, 2014], 60; Parrinder, quoted in ibid., 183).

99. Evan Horowitz, "*Ulysses:* Mired in the Universal," *Modernism/modernity* 13, no. 1 (2006): 878.

100. Leonard Diepeveen, *The Difficulties of Modernism* (New York: Routledge, 2003), 55.

101. George Slocombe's description archly continues, "and with many of the literary and social characteristics of each" (Sam Slote, *"Ulysses" in the Plural: The Variable Editions of Joyce's Novel* [Dublin: National Library of Ireland, 2004], 13). Also see James Joyce, letter to Ezra Pound, October 14, 1916, in Hutton, *Serial Encounters*, 85.

102. In 1932, Odyssey Press used India paper for one of its printings of *Ulysses* (Slote, *"Ulysses" in the Plural,* 18).

103. Stein, *Tender Buttons*, 13, 15, 41.

104. Brian Cosgrove, *James Joyce's Negations: Irony, Indeterminacy and Nihilism in "Ulysses" and Other Writings* (Dublin: University College Dublin Press, 2007), 150.

105. Bloom's name is misprinted as *"L. Boom"* in the newspaper, one of a "crop of nonsensical howlers of misprints" including the meaningless noise ".)eatondph 1/8 ador dorador douradora" (16.1262, 1262–63, 1267, 1257–58). This "line of bitched type" (16.1262–63) was likely Joyce's misrepresentation of the first column of keys on a Linotype machine, "ETAOIN"—akin to hitting a typewriter and accidentally writing "QWERTY" (Hugh Kenner, *The Mechanic Muse* [New York: Oxford University Press, 1987], 8). In the case of *Ulysses,* life imitated art: the first edition contained so many errors that it was issued with a now-famous apology. On Joyce's errors, see Tim Conley, *Joyces Mistakes: Problems of Intention, Irony, and Interpretation* (Toronto: University of Toronto Press, 2003).

106. See Duffy, *The Subaltern "Ulysses";* and Vincent J. Cheng, *Joyce, Race, and Empire* (Cambridge: Cambridge University Press, 1995).

107. I have argued elsewhere that Joyce uses representations of the National Museum's collection in *Ulysses* to suggest that Molly, Stephen, and Bloom are better understood in connection with other subaltern peoples than via representations of colonial Irishness as not-Englishness. See Julia Panko, "Curating the Colony: Museums in *Ulysses,*" *James Joyce Quarterly* 51, nos. 2–3 (2014): 353–70.

108. When Bloom attempts to talk, the Citizen shouts him down: *"Sinn Fein! . . . Sinn fein amhain!"* (12.523). *Sinn féin* means "we ourselves," encapsulating the Citizen's insider-outsider logic.

109. Hope A. Olson, *The Power to Name: Locating the Limits of Subject Representation in Libraries* (Dordrecht, Netherlands: Springer Science and Business Media, 2002), 9.

110. Safiya Umoja Noble discusses the Yellow Peril topic and other examples in *Algorithms of Oppression: How Search Engines Reinforce Racism* (New York: New York University Press, 2018).

111. Edward Said, *Orientalism* (1978; New York: Vintage, 1994), 205.

112. Noble, *Algorithms of Oppression,* 141.

113. See, for example, Alissa Cherry and Keshav Mukunda, "A Case Study in Indigenous Classification: Revisiting and Reviving the Brian Deer Scheme," *Cataloging and Classification Quarterly* 53, nos. 5–6 (2015): 548–67.

114. Olson, *Power to Name,* 234.

115. Institut International de Bibliographie, "III. Les méthodes documentaires," *Publication de l'I.I.B* 106 (1910): 57, 60.

116. Quoted in Duyvis, "Third Report," 13.

117. Wim Van Mierlo, "The Subject Notebook: A Nexus in the Composition History of *Ulysses*—A Preliminary Analysis," *Genetic Joyce Studies* 7 (2007): n.p.

118. Two representative examples are Barbara Stevens Heusel's "Parallax as a Metaphor for the Structure of 'Ulysses'" (*Studies in the Novel* 15, no. 2 [1983]: 135–46) and Hugh Kenner's *Ulysses* (Baltimore: Johns Hopkins University Press, 1980).

For a historically engaged alternative to this tradition of reading parallax formally in Joyce, see Cóilín Parsons, "Planetary Parallax: *Ulysses,* the Stars, and South Africa," *Modernism/modernity* 24, no. 1 (2017): 67–85.

119. See Warren Weaver, "Recent Contributions to the Mathematical Theory of Communication," *ETC* 10, no. 4 (1953), 261–81.

120. I agree with Saint-Amour's argument in *Tense Future* that *encyclopedic* is a better descriptor than *epic* for *Ulysses.* As Saint-Amour writes, modernist encyclopedic novels are more in keeping with the ambivalence, arbitrariness, and volatility that underlay Diderot's *Encyclopédie* than with the totalizing coherence of epic. The question of whether *Ulysses's* complexity can be resolved to reveal an underlying order has been an ongoing debate. For instance, Karen Lawrence argues for the text's irreducible complexity: "*Ulysses* is . . . a compendium of schemes of order that implies there is no absolute way to order experience" (*The Odyssey of Style in "Ulysses"* [Princeton, NJ: Princeton University Press, 1981], 208). Thomas Jackson Rice articulates the contrasting view: in *Ulysses,* he argues, the reader will make "the discovery that chaos is *ordered,* that a vast array of complex and purportedly random phenomena, studied in sufficient detail, reveal deeply embedded patterns" (*Joyce, Chaos, and Complexity* [Urbana: University of Illinois Press, 1997], 84). While I agree that *Ulysses* is interested in complexity and contingency, reading it in the context of information history opens an alternative to reading it as either irreducibly random or entirely explainable once the reader decodes hidden ordering patterns.

121. Michael McKeon, "Genre Theory," in *Theory of the Novel: A Historical Approach,* ed. Michael McKeon (Baltimore: Johns Hopkins University Press, 2000), 4.

122. Woolf, *Mrs. Dalloway,* 21.

123. Ernst Z. Rothkopf, "Incidental Memory for Location of Information in Text," *Journal of Verbal Learning and Verbal Behavior* 10, no. 6 (1971): 608–13; Liang-Yi Li, Gwo-Dong Chen, and Sheng-Jie Yang, "Construction of Cognitive Maps to Improve E-Book Reading and Navigation," *Computers and Education* 60, no. 1 (2013): 32–39; A. Mangen, P. Robinet, G. Oliver, and J. L. Velay, "Mystery Story Reading in Pocket Print Book and On Kindle: Possible Impact on Chronological Events Memory" (presented at a meeting of the International Society for the Empirical Study of Literature and Media, Turin, Italy, July 21–25, 2014).

124. Anne Mangen, quoted in Alison Flood, "Readers Absorb Less on Kindles Than on Paper," *Guardian,* August 19, 2014.

125. J. Paul Hunter, *Before Novels: The Cultural Contexts of Eighteenth-Century Fiction* (New York: Norton, 1990), 29, 30.

126. McKeon, "Genre Theory," 3–4.

127. Hunter, *Before Novels,* 29.

Chapter 2: Form in the Cloud

1. Matthew L. Jockers, *Macroanalysis: Digital Methods and Literary History* (Urbana: University of Illinois Press, 2013), 3, 4.

2. John Unsworth, "Reading at Library-Scale: New Methods, Attention Prosthetics, Evidence, and Argument" (presented at Cyberinfrastructure Days, University of Notre Dame, April 29, 2012); Franco Moretti, *Graphs, Maps, Trees: Abstract Models for Literary History* (London: Verso, 2007), 3, 2.

3. David Nicholas, Ian Rowlands, and Paul Huntington, "Information Behavior of the Researcher of the Future" (presented at University College London, CIBER Group Briefing, for the Joint Information Systems Committee and the British Library, January 11, 2008).

4. Quoted in Angela Chen, "Speed Reading Returns: Apps and Classes Help People Adapt to Reading on Their Phones," *Wall Street Journal*, March 26, 2014.

5. Sven Birkerts, *The Gutenberg Elegies: The Fate of Reading in an Electronic Age*, rev. ed. (New York: Faber and Faber, 2006), 79.

6. Nicholas Carr mentions Charles Dickens's *Nicholas Nickleby* in *The Shallows: What the Internet Is Doing to Our Brains* (London: Norton, 2011), 103. Birkerts discusses Henry James in *The Gutenberg Elegies*, 18. N. Katherine Hayles mentions both Dickens and Jane Austen in "Hyper and Deep Attention: The Generational Divide in Cognitive Modes," *Profession* (2007): 187–88.

7. Adrienne LaFrance, "How Many Websites Are There?," *Atlantic*, September 30, 2015.

8. Frank Pasquale, *The Black Box Society: The Secret Algorithms That Control Money and Information* (Cambridge, MA: Harvard University Press, 2016), 59.

9. Alex Wright, *Glut: Mastering Information through the Ages* (Ithaca: Cornell University Press, 2007), 6.

10. Ralph Jacobson, "2.5 Exabytes (18 Zeroes) of Data Per Day and 80% Is Unstructured," *IBM Consumer Products Industry Blog*, June 24, 2014 (website no longer available).

11. Quoted in Cathy Newman, "Decoding Jeff Jonas, Wizard of Big Data," *National Geographic*, May 6, 2014.

12. Mikal Khoso, "How Much Data Is Produced Every Day?," *Northeastern University: New Ventures*, May 13, 2016.

13. Pasquale, *Black Box Society*, 60.

14. Siva Vaidhyanathan, *The Googlization of Everything (And Why We Should Worry)* (Berkeley: University of California Press, 2011). See also Pasquale, *Black Box Society*; Safiya Umoja Noble, *Algorithms of Oppression: How Search Engines Reinforce Racism* (New York: New York University Press, 2018); and Tarleton Gillespie, "Algorithmically Recognizable: Santorum's Google Problem, and Google's Santorum Problem," *Information, Communication, and Society* 20, no. 1 (2017): 63–80.

15. Noble, *Algorithms of Oppression,* 34.

16. Eric Schmidt, quoted in Randall Stross, *Planet Google: One Company's Audacious Plan to Organize Everything We Know* (New York: Free Press, 2009), 12; Google, quoted in ibid., 9.

17. As early as 1970, E. F. Codd wrote that "future users of large data banks must be protected from having to know how the data is organized in the machine" ("A Relational Model of Data for Large Shared Data Banks," *Communications of the ACM* 13, no. 6 [1970]: 377).

18. Michael Gorman, "The Corruption of Cataloging," *Library Journal* 120, no. 15 (1994): 33.

19. See, for instance, James Gleick, *The Information: A History, a Theory, a Flood* (New York: Pantheon, 2011). Richard Saul Wurman characterized the Internet as "a tsunami of data that is crashing onto the beaches of the civilized world" (*Information Architects* [New York: Graphis, 1997]), 15.

20. Kate Kochetkova, "Seven Amazing Maps of the Internet," *Kaspersky Lab Daily,* November 3, 2015.

21. Alexander R. Galloway, *The Interface Effect* (Cambridge: Polity, 2012), 86.

22. Google, "How Search Works—Organizing Information," *Google Search,* https://www.google.com.

23. Google, "About Search," in Internet Archive, *Wayback Machine,* https://web.archive.org.

24. Craig Dworkin, "Poetry in the Age of Consumer-Generated Content," *Critical Inquiry* 44, no. 4 (2018): 679.

25. Melissa K. Chalmers and Paul N. Edwards, "Producing 'One Vast Index': Google Book Search as an Algorithmic System," *Big Data and Society* 4, no. 2 (2017): 3.

26. Ed Finn, *What Algorithms Want: Imagination in the Age of Computing* (Cambridge, MA: MIT Press, 2017), 20.

27. Google, "About Search."

28. Pasquale, *Black Box Society,* 73.

29. Vaidhyanathan, *Googlization of Everything,* 7.

30. Building on Alexander Galloway's terminology, Ted Striphas defines "algorithmic culture" as "the enfolding of human thought, conduct, organization and expression into the logic of big data and large-scale computation, a move that alters how the category *culture* has long been practiced, experienced, and understood" ("Algorithmic Culture," *European Journal of Cultural Studies* 18, no. 4–5 [2015]: 396).

31. Pasquale, *Black Box Society,* 33–34. Cathy O'Neil, *Weapons of Math Destruction: How Big Data Increases Inequality and Threatens Democracy* (New York: Crown, 2016), 23–27; Motahhare Eslami, Aimee Rickman, Kristen Vaccaro, Amirhossein Aleyasen, Andy Vuong, Karrie Karahalios, Kevin Hamilton, and Christian Sandvig, "'I Always Assumed That I Wasn't Really That Close to [Her]':

Reasoning about Invisible Algorithms in the News Feed," in *Proceedings of the 33rd Annual ACM Conference on Human Factors in Computing Systems, Seoul, Korea, April 2015* (New York: Association for Computing Machinery, 2015), 153–62.

32. See Noble, *Algorithms of Oppression.*
33. Stross, *Planet Google,* 3.
34. Alex Cook, Jonathan Jarvis, and Jonathan Lee, "Evolving the Google Identity," *Google Design,* September 1, 2015, https://design.google.com.
35. Alexander S. Lawson, *Anatomy of a Typeface* (Boston: Godine, 1990), 184.
36. Cook et al., "Evolving the Google Identity."
37. Ibid.
38. Liese Zahabi, *Beyond the Search Engine: Navigating Online Search in the Age of Big Data, Social Media, and Hyperbole* (Interaction Design Foundation, forthcoming).
39. Noble, *Algorithms of Oppression,* 37.
40. Pasquale, *Black Box Society,* 71.
41. Amy Mitchell, Jeffrey Gottfried, Michael Barthel, and Nami Sumida, "Distinguishing between Factual and Opinion Statements in the News," *Journalism .org,* June 19, 2018.
42. See Vaidhyanathan, *Googlization of Everything,* 64–67.
43. Noble, *Algorithms of Oppression,* 111, 148.
44. Eslami et al., "I Always Assumed," 154.
45. Finn, *What Algorithms Want,* 90.
46. Stross, *Planet Google,* 10.
47. Peter Stallybrass, "Books and Scrolls: Navigating the Bible," in *Books and Readers in Early Modern England: Material Studies,* ed. Jennifer Anderson and Elizabeth Sauer (Philadelphia: University of Pennsylvania Press, 2002): 46.
48. Quoted in Joshua Quittner, "Getting Up to Speed on the Computer Highway," *Newsday,* November 3, 1992.
49. Google, "How Search Organizes Information," *Google Search,* https://www .google.com.
50. Chalmers and Edwards, "Producing 'One Vast Index,'" 2.
51. Google, "Google Books Library Project—An Enhanced Card Catalog of the World's Books," *Google Books,* 2012, https://www.google.com.
52. Quoted in Stross, *Planet Google,* 102.
53. Chalmers and Edwards, "Producing 'One Vast Index,'" 6, 3.
54. Simon Peter Rowberry, "Ebookness," *Convergences* 23, no. 3 (2015): 303.
55. Thomas A. Vogler, "When a Book Is Not a Book," in *A Book of the Book: Some Works and Projects about the Book and Writing,* ed. Jerome Rothenberg and Steven Clay (New York: Granary, 2000), 451.
56. Wolfgang Ernst, "Underway to the Dual System: Classical Archives and Digital Memory," trans. Christopher Jenkin-Jones, in *Digital Memory and the Archive,* ed. Jussi Parikka (Minneapolis: University of Minnesota Press, 2013), 84.

57. Chalmers and Edwards, "Producing 'One Vast Index,'" 12.

58. The term is Eli Pariser's; see *The Filter Bubble: What the Internet Is Hiding from You* (New York: Penguin, 2011).

59. Pasquale, *Black Box Society,* 60.

60. Johanna Drucker, "The Virtual Codex from Page Space to E-Space," in *A Companion to Digital Literary Studies,* ed. Ray Siemens and Susan Schreibman (Malden, MA: Blackwell, 2007), 228.

61. Jacques Derrida, "Paper or Me, You Know . . . (New Speculations on a Luxury of the Poor)," in *Paper Machine,* trans. Rachel Bowlby (Stanford: Stanford University Press, 2005), 60; N. Katherine Hayles, *How We Became Posthuman: Virtual Bodies in Cybernetics, Literature, and Informatics* (Chicago: University of Chicago Press, 1999), 47. Elsewhere, Derrida does note that digital environments tend to remediate the book (see, for example, "The Word Processor," in *Paper Machine,* 30). While Hayles's flickering signifier entails not only data stored on a hard drive but also the multiple levels of coding and decoding that intervene between these data and their human reader, it portrays electronic textuality as more unstable and ephemeral than textual inscription on paper is.

62. On material metaphors, see N. Katherine Hayles's *Writing Machines* (Cambridge, MA: MIT Press, 2002), esp. chap. 2.

63. Paul Otlet, "The Union of International Associations: A World Center" (1914), in *International Organisation and Dissemination of Knowledge: Selected Essays of Paul Otlet,* ed. and trans. W. Boyd Rayward (Amsterdam: Elsevier, 1990), 119.

64. Stuart Kelly, "theMystery.doc by Matthew McIntosh Review—A Giant Scrapbook of Ideas," *Guardian,* December 8, 2017; Jason Sheehan, "You're Going to Hate 'theMystery.doc,' and That's OK," *NPR,* October 7, 2017.

65. Stefano Ercolino, *The Maximalist Novel: From Thomas Pynchon's "Gravity's Rainbow" to Roberto Bolaño's "2666"* (New York: Bloomsbury Academic, 2015), xi.

66. John Johnston, *Information Multiplicity: American Fiction in the Age of Media Saturation* (Baltimore: Johns Hopkins University Press, 1998), 2.

67. Tom LeClair, *In the Loop: Don DeLillo and the Systems Novel* (Urbana: University of Illinois Press, 1987); Frederick R. Karl, *American Fictions, 1980–2000: Whose America Is It Anyway?* (Philadelphia: Xlibris, 2001); Hayles, *How We Became Posthuman.*

68. Richard Lea, "The Big Question: Are Books Getting Longer?," *Guardian,* December 10, 2015.

69. Matthew McIntosh, *theMystery.Doc* (New York: Grove, 2017), 385, 391, 438, 439. All ellipses in the quotations are McIntosh's. Dashes in the first entry indicate where McIntosh placed a black rectangle over some text. Further page references appear in the text as parenthetical citations.

70. Kenneth Goldsmith, *Uncreative Writing: Managing Language in the Digital Age* (New York: Columbia University Press, 2011), 31.

71. Dworkin, "Poetry in the Age," 690.

72. Zahabi, *Beyond the Search Engine.*

73. *theMystery.doc* itself is a relatively quick read: although the narrative shifts are disorienting, the prose is fairly conventional. The prose pages, which are often not densely printed, are interspersed with pictures and blank pages.

74. Jessica Pressman, *Digital Modernism: Making It New in New Media* (Oxford: Oxford University Press, 2014), 166.

75. Daniel Punday, *Writing at the Limit: The Novel in the New Media Ecology* (Lincoln: University of Nebraska Press, 2012), 114.

76. Danielewski announced in 2018 that the planned twenty-seven-volume series was on indefinite hiatus.

77. Manuel Portela, *Scripting Reading Motions: The Codex and the Computer as Self-Reflexive Machines* (Cambridge, MA: MIT Press, 2013), 281. In chapter 5, Portela documents *Only Revolutions*'s many symmetries and correspondences.

78. N. Katherine Hayles, *How We Think: Digital Media and Contemporary Technogenesis* (Chicago: University of Chicago Press, 2012), 230.

79. Richard Jozsa and Stuart Presnell, "Universal Quantum Information Compression and Degrees of Prior Knowledge," *Proceedings* 459, no. 2040 (2003): 3061.

80. Gleick, *The Information,* 344.

81. Jonathan Sterne, *MP3: The Meaning of a Format* (Durham, NC: Duke University Press, 2012), 1–2.

82. I. A. Richards, *Principles of Literary Criticism* (1926; New York: Routledge, 2001), 274.

83. For example, Jacques Roubad argues that a poem's ability to condense experience into language is closely tied to its storage potential, its "memory": "In a poem, what springs from poetry's memory, by the arrangement of language, is a state of compression, of condensation, of compactification" (Quoted in Agnès Disson and Roxanne Lapidus, "Pierre Alferi: Compressing and Disconnecting" *Substance* 39, no. 3 [2010]: 78).

84. Mark Z. Danielewski, "The Futures of the Book" (lecture at University of California, Santa Barbara, February 24, 2011).

85. Joe Bray, "Going in Circles: The Experience of Reading *Only Revolutions,*" in *Revolutionary Leaves: The Fiction of Mark Z. Danielewski,* ed. Sascha Pöhlmann (Newcastle upon Tyne, UK: Cambridge Scholars Publishing, 2012), 183.

86. Mark Z. Danielewski, *Only Revolutions* (New York: Pantheon, 2006). Further page references appear in the text as parenthetical citations.

87. "The Presidency: The Government Still Lives," *Time,* November 29, 1963.

88. Hayles, *How We Think,* 223.

89. As Pressman notes, "the Creep seems to represent normative values and institutionalized roles" (*Digital Modernism,* 163).

90. Sascha Pöhlmann, "The Democracy of Two: Whitmanian Politics in *Only Revolutions*," in Pöhlmann, *Revolutionary Leaves*, 23.

91. See the "Road Sign" annotations for page 252 of Sam's narrative and page 16 of Hailey's narrative (Julia Panko and Noam Assayag, "'Road Sign' Annotations," in Mark Z. Danielewski, *Only Revolutions: The Interactive EBook* [New York: Penguin Random House/iBooks, 2015]).

92. Although, according to Danielewski, "no technology appears in [*Only Revolutions*]," the Internet is covertly present in the novel, evident in the RSS feed–like structure of the chronomosaics as well as its companion website (which prominently featured an RSS feed with links in the style of the chronomosaics) (Callie Miller and Michele Reverte, "LAist Interview: Mark Danielewski," *LAist*, October 23, 2007).

93. Mark Z. Danielewski, personal communication. I am grateful to him for describing his research process to me.

94. Anne Mangen and Jean-Luc Velay write that "recent theoretical currents in psychology, phenomenology & philosophy of mind, and neuroscience—commonly referred to as 'embodied cognition'—indicate that perception and motor action are closely connected and, indeed, reciprocally dependent" ("Digitizing Literacy: Reflections on the Haptics of Writing," in *Advances in Haptics*, ed. Mehrdad Hosseini Zadeh [InTechOpen, 2010], 385); Anne Mangen, Bente R. Walgermo, and Kolbjørn Brønnick, "Reading Linear Texts on Paper versus Computer Screen: Effects on Reading Comprehension," *International Journal of Educational Research* 58, no. 1 (2013): 66.

95. Hayles, *How We Think*, 231.

96. Danielewski, personal communication.

97. The phrase also acknowledges readers' contributions to the novel: "In writing *Only Revolutions*, Danielewski is said to have called upon dedicated readers and fans through online discussion boards, asking for responses to famous events and for details of their favourite animals, many of which were used in the book. The book itself therefore does contain within it the personal touches of some of its readers, adding another dimension to the second-person pronoun of 'You were there'" (Alison Gibbons, "'You Were There': The Allways Ontologies of *Only Revolutions*," in Pöhlmann, *Revolutionary Leaves*, 171).

98. Lisa Nakamura, "'Words with Friends: Socially Networked Reading on Goodreads," *PMLA* 128, no. 1 (2013): 239.

99. Lori Emerson, *Reading Writing Interfaces: From the Digital to the Bookbound* (Minneapolis: University of Minnesota Press, 2014).

100. Robert Fitterman, "Foreword," in *Notes on Conceptualisms*, ed. Vanessa Place and Robert Fitterman (Brooklyn: Ugly Duckling Presse, 2009), 10.

101. For instance, Fitterman argues that conceptual writing has primarily "taken hold in the poetry community" (ibid., 9).

102. Goldsmith, *Uncreative Writing,* 5, 1.

103. Walter Benjamin, "The Task of the Translator: An Introduction to the Translation of Baudelaire's *Tableaux parisiens*" (1923), in *Illuminations: Essays and Reflections,* ed. Hannah Arendt, trans. Harry Zohn (New York: Schocken, 1968), 69; Ludwig Wittgenstein, *Zettel,* ed. G. E. M. Anscombe and G. H. von Wright, trans. G. E. M. Anscombe (Berkeley: University of California Press, 1967), 28e.

104. For example, Jerome McGann argues that an imaginative work "is organized as rhetoric and *poiesis* rather than as exposition and information-transmission" (*Radiant Textuality: Literature after the World Wide Web* [New York: Palgrave, 2001], 33); and William Paulson, whose *The Noise of Culture* is subtitled "Literary Texts in a World of Information," maintains that "we do not need all of the paraphrasable information contained in the literature we read" (*The Noise of Culture: Literary Texts in a World of Information* [Ithaca: Cornell University Press, 1988], 99).

105. J. Paul Hunter, *Before Novels: The Cultural Contexts of Eighteenth-Century Fiction* (New York: Norton, 1992), 52, 53.

106. Jockers, *Macroanalysis,* 8–9.

107. Mikhail Mikhaĭlovich Bakhtin, *The Dialogic Imagination* (1981), ed. Michael Holquist, trans. Caryl Emerson and Michael Holquist (Austin: University of Texas Press, 2004), 39.

108. On the poetic dimensions of *Only Revolutions,* see Brian McHale, "*Only Revolutions,* or, the Most Typical Poem in World Literature" in *Mark Z. Danielewski,* ed. Joe Bray and Alison Gibbons (Manchester, UK: Manchester University Press, 2011), 141–58.

109. Danielewski, "The Futures of the Book."

110. Goldsmith, *Uncreative Writing,* 10.

111. Julia Flanders, "Detailism, Digital Texts, and the Problem of Pedantry," *Text Technology* 14, no. 2 (2005): 54.

112. Ibid.

113. Karen Reimer, untitled post, *Book Arts,* https://www.bookarts.uwe.ac.uk.

114. Karen Reimer [as Eve Rhymer], *Legendary, Lexical, Loquacious Love* (Chicago: Sara Ranchouse, 1996), 22.

115. Dworkin, "Poetry in the Age," 684.

116. For a comprehensive reading list in this area, see Tarleton Gillespie and Nick Seaver's "Critical Algorithm Studies: A Reading List," *SocialMediaCollective.org,* December 15, 2016.

117. Lisa Gitelman, *Always Already New: Media, History, and the Data of Culture* (Cambridge, MA: MIT Press, 2000), 86.

118. See Sterne, *MP3,* chap. 5.

119. See Noble, *Algorithms of Oppression.*

120. Kathleen Fitzpatrick, "The Exhaustion of Literature: Novels, Computers, and the Threat of Obsolescence," *Contemporary Literature* 43, no. 3 (2002): 525n. Fitzpatrick argues that although the anxiety that novels will become obsolete in the wake of new technologies "is not confined to white male writers, it is largely aligned with a white male subject position" (ibid.). On the dominance of this perspective in information history and the development of digital tools, see Noble, *Algorithms of Oppression;* and Emily Chang, *Brotopia: Breaking Up the Boys' Club of Silicon Valley* (New York: Penguin, 2018).

121. Where *Only Revolutions* champions the marginalized, *theMystery.doc* is more ambivalent in its treatment of minority groups. It is difficult to determine whether McIntosh's casual accounts of events such as domestic abuse and gang violence are meant to be taken at face value or read as ironic critique. The novel does, however, raise the question of how skeptical the reader should be of the framing narrator's perspective, which often descends into stereotyping or affectless apathy.

122. Goldsmith, *Uncreative Writing,* 123–24.

123. Cathy Park Hong, "There's a New Movement in American Poetry and It's Not Kenneth Goldsmith," *New Republic,* October 1, 2015.

124. Sara McLafferty, "Women and GIS: Geospatial Technologies and Feminist Geographies," *Cartographica* 40, no. 4 (2005): 38.

Chapter 3: Haptic Storage

1. Edison's predictions were published in the June 20, 1911, edition of the *Miami Metropolis.* For a reprint, see Matt Novak, "Edison's Predictions for the Year 2011 (1911)," *Paleofuture,* January 18, 2011, http://gizmodo.com.

2. Paul Otlet and Robert Goldschmidt, "On a New Form of the Book: The Microphotographic Book," in *International Organisation and Dissemination of Knowledge: Selected Essays of Paul Otlet,* trans. and ed. W. Boyd Rayward (Amsterdam: Elsevier, 1990), 89.

3. William Saffady, *Micrographics* (Littleton, CO: Libraries Unlimited, 1978), 13.

4. The most widely read work on microform published in the past two decades is certainly Nicholson Baker's anti-microfilm treatise *Double Fold: Libraries and the Assault on Paper* (New York: Random House, 2001).

5. Leah Price, "Reading as if for Life," *Michigan Quarterly Review* 48, no. 4 (2009): 483.

6. Jonathan Auerbach and Lisa Gitelman, "Microfilm, Containment, and the Cold War," *American Literary History* 19, no. 3 (2007): 747.

7. M. H. Garbanati, "Micro-Photography," *Photographic News,* February 4, 1859, 262.

8. That microphotography could be advantageous for information storage was demonstrated in the 1870s, when besieged forces in Paris sent microphotographed documents via carrier pigeons during the Franco-Prussian War. It took several

more decades, however, for technological developments to facilitate microform's adoption in textual storage.

9. Otlet and Goldschmidt, "On a New Form of the Book," 87–88, 89.

10. Paul Otlet and Robert Goldschmidt, "The Preservation and International Diffusion of Thought: The Microphotic Book" (1925), in Rayward, *International Organisation and Dissemination of Knowledge,* 89, 205.

11. Otlet and Goldschmidt, "On a New Form of the Book," 93.

12. Michael Buckland, "On the Cultural and Intellectual Context of European Documentation in the Early Twentieth Century," in *European Modernism and the Information Society: Informing the Present, Understanding the Past,* ed. W. Boyd Rayward (Aldershot, UK: Ashgate, 2008), 50.

13. Quoted in Alan Marshall Meckler, *Micropublishing: A History of Scholarly Micropublishing in America, 1938–1980* (Westport: Greenwood, 1982), 43.

14. Edward A. Henry, "Books on Film: Their Use and Care," *Library Journal* 57 (1932): 217.

15. David Thorburn and Henry Jenkins, "Introduction: Towards an Aesthetics of Transition," in *Rethinking Media Change: The Aesthetics of Transition,* ed. David Thorburn and Henry Jenkins (Cambridge, MA: MIT Press, 2003), 1–2.

16. Harry Miller Lydenberg, "Should the 'Times,' 1914–1918 Be Filmed?" (1935), in *Studies in Micropublishing, 1853–1976,* ed. Allen B. Veaner (Westport, CT: Microform Review, 1976), 403.

17. M. Llewellyn Raney, speech at the American Library Association's Microphotography Symposium (1936), quoted in Baker, *Double Fold,* 73; Robert C. Binkley, quoted in Allen B. Veaner, "Introduction to Section X," in Veaner, *Studies in Micropublishing,* 441. The "series of changes" Binkley mentions refers to photo-offset printing and near-print as well as microfilm.

18. Meckler, *Micropublishing,* 154.

19. Eugene Kinkead, "The Admiral's Chair," *New Yorker,* February 7, 1942, 11. Fiske invented the telescopic gunboat sights that form the basis for the modern rangefinder. His other inventions included "a boat detaching apparatus, a system for electric communication for the interior of warships, a stadimeter," and "a battle-order telegraph" (Frank J. Costello, "Urges Books of Tiny Type," *Brooklyn Daily Eagle,* April 3, 1927).

20. "Admiral Fiske's New Invention," *New York Times,* March 5, 1922.

21. "Reading Machine Invented," *Library Journal* 57 (March 1932): 249.

22. S. R. Winters, "Stretching the Five-Foot Shelf," *Scientific American* (June 1922): 407.

23. "A Dictionary in Your Vest Pocket," *Literary Digest,* June 10, 1922, 25.

24. Otlet and Goldschmidt, "The Preservation and International Diffusion of Thought," 205.

25. "Reading Machine to Aid Eyes and Reduce Cost of Books," *Popular Mechanics* 46, no. 1 (1926): 108.

26. Ibid. A miniaturized microform phonebook was produced to this end in 1928, reducing the directory to three-quarters of an inch in thickness. Susan Stewart evocatively describes the project: "In [Fiske's] glass the eight million stories of the Naked City opened into an accordion of significance" (*On Longing: Narratives of the Miniature, the Gigantic, the Souvenir, the Collection* [Durham, NC: Duke University Press, 1993], 41).

27. "Admiral Fiske Invents Pocket Bookcase," *Popular Science Monthly* 100, no. 6 (1922): 48; Costello, "Urges Books of Tiny Type," 3. Fiske estimated that his machine would allow readers to carry "one hundred thousand words, the length of an average book, on a tape slightly longer than forty inches" (Bradley A. Fiske, reading machine, U.S. Patent 1,411,008, filed November 20, 1920, issued March 28, 1922).

28. "New Device to Replace Printing with Photo Engraving," *American Photo Engraver* 14, no. 5 (1922): 216.

29. "Reading Machine to Aid Eyes," 108.

30. "How Science Makes Libraries Pocket-Size," *Tennessean,* March 23, 1941.

31. Bob Brown, *The Readies* (Bad Ems, Germany: Roving Eye Press, 1930), 1. Additional references to this work will appear parenthetically in the text.

32. Bob Brown, "Appendix," in *Readies for Bob Brown's Machine* (Cagnes-sur-Mer, France: Roving Eye Press, 1931), 172. Additional references to this work will appear parenthetically in the text. In an unpublished, undated manuscript titled "Plans for Study," Brown writes that he "conceived the idea of reading by machine in 1916" (box 32, folder "Reading Machine—Notes, Correspondence, Clippings," Bob Brown Papers, Charles E. Young Research Library, University of California, Los Angeles).

33. Montparno, "Left Bankers Believe Bob Brown's Pill Box Book Reading Machine Will Help Them Absorb Dozen Gertrude Stein Novels in Afternoon," *Chicago Tribune,* January 13, 1930.

34. "How Science Makes Libraries Pocket-Size," 46. The prototype was built by Ross Saunders, who was assisted by Hilaire Hiler (Craig Saper, *The Amazing Adventures of Bob Brown* [New York: Fordham University Press, 2016], 163).

35. Bob Brown, letter to Gertrude Stein, October 8, 1930, Series II: Correspondence of Gertrude Stein—General Correspondence, box 99, folder 1915, Gertrude Stein and Alice B. Toklas Papers, Beinecke Rare Book and Manuscript Library, Yale University.

36. Bob Brown, letter to Gertrude Stein, September 23 [1929 or 1930], Series II: Correspondence of Gertrude Stein—General Correspondence, box 99, folder 1915, Gertrude Stein and Alice B. Toklas Papers.

37. Bob Brown, letter to Gertrude Stein, December 16, 1931, box 32, folder "Reading Machine—Correspondence Includes Mention of Gertrude Stein," Bob Brown Papers.

38. Saper, *Amazing Adventures,* 1, 2.

39. Michael North, "Words in Motion: The Movies, the Readies, and the 'Revolution of the Word,'" *Modernism/modernity* 9, no. 2 (2002): 215.

40. Bob Brown, letter to Hayka n Texhnka, undated (between 1930 and 1932), box 32, folder "Reading Machine—Notes, Correspondence, Clippings," Bob Brown Papers.

41. Earlier influential scholarship on Brown and the Readies includes Jerome McGann's *Black Riders: The Visible Language of Modernism* (Princeton, NJ: Princeton University Press, 1993); and Craig Dworkin's "'Seeing Words Machinewise': Technology and Visual Prosody," *Sagetrieb* 18 (Spring 1999): 59–86. Saper has been the most enthusiastic and prolific chronicler of Brown's life and work, and I am indebted to his efforts in making Brown's texts more widely available. In addition to writing an engaging biography of Brown, he has edited new editions of a number of Brown's books and has built an online digital simulation of the Readies machine (http://www.readies.org/).

42. On the Readies and cinema, see North, "Words in Motion"; on the tachistoscope, see Jessica Pressman, "Machine Poetics and Reading Machines: William Poundstone's Electronic Literature and Bob Brown's Readies," *American Literary History* 23, no. 4 (2011): 767–94. Pressman and North also draw connections between Brown's experiment and digital reading. North notes that Brown's machine "appears to have anticipated the microfilm reader" ("Words in Motion," 215), and Dworkin describes the machine as "something like a cross between the recently invented microfilm reader and Microsoft Word" ("Seeing Words Machinewise," 59). Neither, however, develops this connection further or discusses Brown's conceptual debt to the discourse of microform innovation. Saper describes the Readies machine as "resembl[ing] a microfilm reader" and briefly mentions Brown's connection to Fiske, but he focuses on the differences between them (*Amazing Adventures of Bob Brown,* 160).

43. Bob Brown, letter to Bradley A. Fiske, January 5, 1932, box 32, folder "Reading Machine—correspondence, clippings, 1930 Sept. 24–1932 Feb. 11; 1941 March 23," Bob Brown Papers.

44. Saper, *Amazing Adventures,* 184.

45. Brown, *Readies,* 29. The eye is a dominant motif in Brown's oeuvre, from the name of his Roving Eye Press to his description of the Readies as "the Optical Art of Writing" (27).

46. Stephens, *Poetics of Information Overload,* 74.

47. While microform advocates frequently dismissed the print book as "cumbersome," this complaint also surfaced in many reactions to microform readers, and it prevented their widespread adoption. On this topic, and on the implications of microform's failures for media history, see Julia Panko, "'A New Form of the Book': Modernism's Textual Culture and the Microform Moment," *Book History* 22 (2019): 342–69.

48. See North, "Words in Motion."

49. Matthew G. Kirschenbaum, *Mechanisms: New Media and the Forensic Imagination* (Cambridge, MA: MIT Press, 2008), 39.

50. See Auerbach and Gitelman, "Microfilm, Containment, and the Cold War."

51. See McGann's *Black Riders;* and Johanna Drucker, *The Visible Word: Experimental Typography and Modern Art, 1909–1923* (Chicago: University of Chicago Press, 1994), 60.

52. Filippo Marinetti, "The Destruction of Syntax—Imagination without Strings—Words-in-Freedom" (1913), in *A Book of the Book: Some Works and Projections about the Book and Writing,* ed. Jerome Rothenberg and Steven Clay (New York: Granary, 2000), 183.

53. In Beatrice Warde's view, most famously expressed in her essay "The Crystal Goblet, or Printing Should Be Invisible" (first given as a talk to the British Typographers Guild in 1932), the printed page functioned best as a seamless conduit "between the reader . . . and that landscape which is the author's words" (in *The Crystal Goblet: Sixteen Episodes on Typography,* ed. Henry Jacob [Cleveland: World, 1956], 15). As the connoisseur of fine wines prefers a crystal goblet to an ornate chalice because the latter obscures the beauty of the wine it contains, so the connoisseur of fine words prefers "the transparent page" to one on which elaborate typography competes with the meaning of the words (17).

54. Dworkin, "Seeing Words Machinewise," 51.

55. Drucker, *The Visible Word,* 95.

56. Donna E. Rhein, *The Handprinted Books of Leonard and Virginia Woolf at the Hogarth Press, 1917–1932* (Ann Arbor: UMI Research Press, 1985), 42–43.

57. Dworkin, "Seeing Words Machinewise," 36.

58. Lawrence Rainey, *Institutions of Modernism: Literary Elites and Public Culture* (New Haven, CT: Yale University Press, 1988), 67.

59. Quoted in John Lehmann, *Thrown to the Woolfs* (London: Weidenfeld and Nicolson, 1978), 27.

60. Victoria L. Smith, "'Ransacking the Language': Finding the Missing Goods in Virginia Woolf's 'Orlando,'" *Journal of Modern Literature* 29, no. 4 (2006): 69.

61. Virginia Woolf, *The Diary of Virginia Woolf,* vol. 3, *1925–1930,* ed. Anne Oliver Bell (London: Hogarth, 1980), 168; Elizabeth Cooley, "Revolutionizing Biography: 'Orlando,' 'Roger Fry,' and the Tradition," *South Atlantic Review* 55, no. 2 (1990): 73. The tonal variation may stem partly from Woolf's ambiguous feelings about her sexual identity. Adam Parkes argues that "[her] well-known vacillation between the comic and the serious while writing *Orlando*" was "a vacillation closely bound up with her uncertain attitude to 'sapphism'" ("Lesbianism, History, and Censorship: The Well of Loneliness and the Suppressed Randiness of Virginia Woolf's *Orlando,*" *Twentieth Century Literature* 40, no. 4 [1994]: 446).

62. Cooley, "Revolutionizing Biography," 79.

63. The character Orlando is on the path to achieving actual immortality. The novel runs from 1553 to 1928; in this span of nearly four centuries, Orlando has aged only twenty years.

64. Woolf, quoted in Victoria [Vita] Sackville-West, *The Letters of Vita Sackville-West to Virginia Woolf,* ed. Louise DeSalvo and Mitchell A. Leaska (London: Hutchinson, 1984), 279; Sackville-West, in ibid., 281–82.

65. Mary Poovey, *Making a Social Body: British Cultural Formation, 1830–1864* (Chicago: University of Chicago Press, 1995), 4.

66. Cornelia Vismann, *Files: Law and Media Technology,* trans. Geoffrey Winthrop-Young (Stanford, CA: Stanford University Press, 2008), 117.

67. R. Buckminster Fuller, *Critical Path* (New York: St. Martin's Press, 1981), 137.

68. Ibid., 128–29, 134.

69. "Guide to the R. Buckminster Fuller Papers M1090," *Online Archive of California,* http://oac.cdlib.org.

70. Fuller, *Critical Path,* 124.

71. As I wrote in the introduction, facts are defined as self-evident statements of reality, whereas "data are representations . . . used as evidence" (Christine L. Borgman, *Big Data, Little Data, No Data: Scholarship in the Networked World* [Cambridge, MA: MIT Press, 2015], 28). When details about an individual's life are contextualized within representational projects such as a biography, they become data—givens that are relative to a construction of an individual rather than independent, objective facts.

72. Gertrude Stein, quoted in Paul Stephens, *The Poetics of Information Overload* (Minneapolis: University of Minnesota Press 2015), 54; Stephens, ibid.

73. Virginia Woolf, *Orlando: A Biography* (1928), ed. Mark Hussey (New York: Harcourt, 2006), 49. Additional references to this work will appear parenthetically in the text. Some of *Orlando*'s lacunae were ways to avoid censorship. Radclyffe Hall landed in legal difficulties for her portrayal of lesbianism in *The Well of Loneliness,* and Parkes argues that the gaps in *Orlando* are the result of "an author buckl[ing] under the pressure of social forces—the forces converging in the censorship of works that the authorities considered obscene" ("Lesbianism, History, and Censorship," 457). While I agree that Woolf glosses over the portions of Orlando's life that lend themselves most readily to accusations of homosexuality, these omissions also fit within *Orlando*'s larger thematic problematizing of representation.

74. Virginia Woolf, "The New Biography," in *Selected Essays,* ed. David Bradshaw (Oxford: Oxford University Press, 2008), 229, 100.

75. *Census Returns of England and Wales, 1911* (Kew: National Archives of the United Kingdom, 1911); "Adeline Virginia Stephen" [image], *Ancestry.com.* At the time of this writing, census records post-dating the 1911 census are not available digitally.

76. For consistency, I have referred to Orlando using the pronoun "her" except when I quote others' language.

77. Hermione Lee, *Virginia Woolf* (London: Chatto and Windus, 1996), 522.

78. Kevin Birmingham, *The Most Dangerous Book: The Battle for James Joyce's "Ulysses"* (New York: Penguin Random House, 2015), 182.

79. Patrick Collier, "Virginia Woolf in the Pay of Booksellers: Commerce, Privacy, Professionalism, *Orlando*," *Twentieth-Century Literature* 48, no. 4 (2002): 376.

80. Woolf, "The New Biography," 233.

81. Woolf's father was famous for his biographical work. For a fuller consideration of Woolf's relationship to biography, see Cooley's "Revolutionizing Biography"; and Catherine Neal Parke's *Biography: Writing Lives* (New York: Routledge, 2002), chap. 3.

82. In addition to its text, *Orlando* includes a mix of photographs and reproduced paintings that purport to show Orlando and her lovers. Woolf's inclusion of this material reminds the reader that the book is a multilayered storage system, capable of incorporating different media. Yet pictorial representation is also fraught in *Orlando*: images are no more able to fully preserve their subjects than the written word is. The three photographs of Vita confuse the real woman with the fictional character. The correspondences between Vita and Orlando, and Vita and her own image, are further challenged because paintings of two of Vita's ancestors are also captioned "Orlando." Nor do the images adhere evenly to the conventions of *roman à clef*: while several photographs picture their real-life antecedents, others do not. These choices throw the ontological status of the novel's images into doubt. As Helen Wussow argues, Woolf uses images in *Orlando*, as she would later in *Flush: A Biography*, *Three Guineas*, and *Roger Fry: A Biography*, "to call into question [the] factuality [of the images] and the overall stability of the photographic subject/object" ("Virginia Woolf and the Problematic Nature of the Photographic Image," *Twentieth Century Literature* 40, no. 1 [1994]: 2).

83. Laura U. Marks, *Touch: Sensuous Theory and Multisensory Media* (Minneapolis: University of Minnesota Press, 2002), viii.

84. Leonard Woolf, *Beginning Again: An Autobiography of the Years 1911–1918* (London: Hogarth Press, 1964), 233.

85. "How my hand writing goes down hill! Another sacrifice to the Hogarth Press" (Woolf, *Diary*, 3:42).

86. Nancy Cunard, *These Were the Hours: Memories of My Hours Press, Réanville and Paris, 1928–1931*, ed. Hugh Ford (Carbondale: Southern Illinois University Press, 1969), 8.

87. Marcus attributes the influence to Woolf's typesetting of Eliot's poetry ("Virginia Woolf and the Hogarth Press," 132); Julia Briggs cites her typesetting of Hope Mirrlees's *Paris* (*Reading Virginia Woolf* [Edinburgh: Edinburgh University Press, 2006], 91).

88. See John Lurz, *The Death of the Book: Modernist Novels and the Time of Reading* (New York: Fordham University Press, 2016), chaps. 4 and 5.

89. Virginia Woolf, *Jacob's Room* (1922; reprint, London: Hogarth, 1990), 33; Virginia Woolf, *The Waves* (1931), ed. James M. Haule and Philip H. Smith, Jr. (Oxford: Blackwell, 1993), 136.

90. Thomas S. W. Lewis, "Virginia Woolf's Sense of the Past," *Salmagundi* 68/69 (1985–86): 193.

91. Elizabeth Freeman, *Time Binds: Queer Temporalities, Queer Histories* (Durham, NC: Duke University Press, 2010), 95, 109, 110. When I proposed the term *haptic storage* in an article (Julia Panko, "'Memory Pressed Flat Into Text': The Importance of Print in Steven Hall's *The Raw Shark Texts*," *Contemporary Literature* 52, no. 2 [2011]: 264–97), I was unaware of Freeman's text, in which she (also inspired by Laura Marks) termed her method *haptic historiography* (123). While our approaches share similarities, I focus on physical preservation as a one mode of storage within a wider configuration of practices and technologies of information storage, whereas Freeman explicates the emotional and sexual aspects of encounters with the past.

92. Madeline Moore, "Orlando: An Edition of the Manuscript," *Twentieth Century Literature* 25, nos. 3–4 (1979): 342n.

93. Charles Norris Williamson and Alice Muriel Williamson's *The Heather Moon* ([Garden City, NY: Doubleday, 1912], 2010) and John Murray's *Handbook for Travellers in Scotland, 1903* ([London: Stanford, 1903], 108) both place Queen Mary's prayer book at Terregles House in Scotland near the time of *Orlando*'s composition.

94. Woolf, *Diary*, 3:125.

95. Amaranth Borsuk, *The Book* (Cambridge, MA: MIT Press, 2018), 77.

96. Allison Muri, "Virtually Human: The Electronic Page, the Archived Body, and Human Identity," in *The Future of the Page,* ed. Peter Stoicheff and Andrew Taylor (Toronto: University of Toronto Press, 2004), 237, 235. Muri argues that "the analogies between page and body are ancient" (236).

97. Peter Stoicheff and Andrew Taylor, "Introduction: Architectures, Ideologies, and Materials of the Page," in Stoicheff and Taylor, *The Future of the Page,* 5–6. As Bruce Holsinger notes, the history of parchment is also a history of the pain of animals—a category of bodily suffering often overlooked in allegories relating books and writing to human bodies. "In parchment cultures," he writes, "the animal never is simply the object of representation but constitutes the material substance of the literary object" ("Of Pigs and Parchment: Medieval Studies and the Coming of the Animal," *PMLA* 124, no. 2 [2009]: 619).

98. Walter Benjamin, "The Work of Art in the Age of Mechanical Reproduction," in *Illuminations: Essays and Reflections,* ed. Hannah Arendt, trans. Harry Zohn (New York: Schocken, 1968), 221.

99. Ibid.

100. Quoted in Sackville-West, *Letters*, 254.

101. Benjamin, "The Work of Art," 221.

102. Walter Benjamin, "Unpacking My Library" (1931), in *Illuminations*, 61.

103. Stewart, *On Longing*, 34.

104. Benjamin, "Unpacking My Library," 67.

105. H. J. Jackson, *Marginalia: Readers Writing in Books* (New Haven, CT: Yale University Press, 2001), 185.

106. Lee, *Virginia Woolf*, 520.

107. Printing also created a social network for the Woolfs, who published the work of their friends. The Hogarth Press's network linked Woolf to Bob Brown, albeit indirectly: Nancy Cunard, Brown's some-time publisher, named her Hours Press after Woolf's working title for *Mrs. Dalloway.*

108. Collier, "Virginia Woolf," 365.

109. Karyn Z. Sproles, "Virginia Woolf Writes to Vita Sackville-West (and Receives a Reply): *Aphra Behn, Orlando, Saint Joan of Arc,* and Revolutionary Biography," in *Virginia Woolf: Texts and Contexts,* ed. Beth Rigel Daugherty and Eileen Barrett (New York: Pace University Press, 1996), 189.

110. Rhein, *Handprinted Books,* 81, 82. Woolf took great care in printing *Sissinghurst,* which was "one of the most perfectly designed and executed of all the [Woolfs'] handprinted books" (30).

111. Sackville-West, *Letters,* 238.

112. Lee, *Virginia Woolf,* 487; Moore, "Orlando: An Edition of the Manuscript," 305.

113. On the relationship between Christianity's "miracle of incarnation and the carnal properties of the book," see John T. Hamilton, "Pagina Abscondita: Reading in the Book's Wake," in *Book Presence in a Digital Age,* ed. Kiene Brillenburg Wurth, Kári Driscoll, and Jessica Pressman (New York: Bloomsbury Academic, 2018), 40.

114. Cunard, *These Were the Hours,* 9.

115. Quoted in Isabelle Rieusset-Lemarié, "P. Otlet's Mundaneum and the International Perspective in the History of Documentation and Information Science," in *Historical Studies in Information Science,* ed. Trudi Bellardo Hahn and Michael Keeble Buckland (Medford, NJ: Information Today, 1998), 38.

116. Bradley Allen Fiske, *From Midshipman to Rear-Admiral* (New York: Century, 1919), 592.

117. Bob Brown, *1450–1950* (New York: Jargon, 1959), 3.

118. The "fly speck" is polysemous; in addition to a mark left on the text, it may refer to the leaf-like symbol repeated several times in the poem. Cunard called the microscopic type in Brown's later publication, *Words,* "fly-speck letters" (*These Were the Hours,* 134).

Chapter 4: Bodies of Information

1. The designation is somewhat dubious, overlooking notable examples of first-generation hypertext novels such as Judy Malloy's *Uncle Roger* (1987) and Michael Joyce's *afternoon: a story* (1990).

2. Peter James, quoted in Alison Flood, "Where Did the Story of EBooks Begin?," *Guardian,* March 12, 2014.

3. Ibid.

4. Quoted in Sarah Shaffi, "Science Museum to Display James Novel," *Bookseller,* March 6, 2014.

5. One notable exception is Hanson Robotics's "Philip K. Dick Android," first activated in 2005. When asked questions, it responds using its synthesis of the author's writings (https://www.hansonrobotics.com/).

6. Luka, "Replika," 2018 (https://replika.ai).

7. Arielle Pardes, "The Emotional Chatbots Are Here to Probe Our Feelings," *Wired,* January 31, 2018.

8. Kuyda, quoted in Casey Newton, "Speak, Memory," *Verge,* October 7, 2016.

9. Quoted in Dan Tynan, "Augmented Eternity: Scientists Aim to Let Us Speak from Beyond the Grave," *Guardian,* June 23, 2016.

10. Frank Pasquale, *The Black Box Society: The Secret Algorithms That Control Money and Information* (Cambridge, MA: Harvard University Press, 2016), 4.

11. Victoria Vesna, "Seeing the World in a Grain of Sand: The Database Aesthetic of Everything," in *Database Aesthetics: Art in the Age of Information,* ed. Victoria Vesna (Minneapolis: University of Minnesota Press, 2007), 8.

12. Quoted in Tynan, "Augmented Eternity," brackets in original.

13. Calculations based on information in LexisNexis, "How Many Pages in a Gigabyte?," *LexisNexis,* 2007, https://www.lexisnexis.com.

14. Rita Raley, "Dataveillance and Countervailance," in *"Raw Data" Is an Oxymoron,* ed. Lisa Gitelman (Cambridge, MA: MIT Press, 2013), 124.

15. Ibid.

16. Bell is a researcher emeritus with Microsoft. He describes the duration of MyLifeBits as "c1998–2007," though he and Gemmell published their book on MyLifeBits in 2009 (https://gordonbell.azurewebsites.net).

17. Microsoft, "MyLifeBits," *Microsoft Research,* https://www.microsoft.com.

18. Gordon Bell and Jim Gemmell, *Total Recall: How the E-Memory Revolution Will Change Everything* (New York: Dutton, 2009), 50. Although *Total Recall,* which describes MyLifeBits, was co-authored, the majority of the book is written as Bell's first-person narration. I thus attribute many of the quotes from *Total Recall* directly to him when I describe them in the text. The book was later reissued under the title *Your Life, Uploaded: The Digital Way to Better Memory, Health, and Productivity.*

19. Ibid., 6.

20. Ibid., 180–81, 73, 187.

21. N. Katherine Hayles, *How We Became Posthuman: Virtual Bodies in Cybernetics, Literature, and Informatics* (Chicago: University of Chicago Press, 1999), 2.

22. Ibid., 2–3.

23. Ted Striphas, "Algorithmic Culture," *European Journal of Cultural Studies* 18, nos. 4–5 (2015): 399.

24. Matthew G. Kirschenbaum, *Mechanisms: New Media and the Forensic Imagination* (Cambridge, MA: MIT Press, 2008), 17.

25. Nick Montfort proposed the term *screen essentialism* in 2004 ("Continuous Paper: The Early Materiality and Workings of Electronic Literature" [presented at a meeting of the Modern Language Association, Philadelphia, December 28, 2004]).

26. Ed Finn, *What Algorithms Want: Imagination in the Age of Computing* (Cambridge, MA: MIT Press, 2017), 62.

27. Ibid., 25.

28. Nick Bostrom, "A History of Transhumanist Thought," *Journal of Evolution and Technology* 14, no. 1 (2005): n.p.

29. Hayles, *How We Became Posthuman,* xiv.

30. Association of American Publishers, "BookStats Publishing Formats Highlights," *Publishers.org,* 2012.

31. Claire Cain Miller, "E-Books Top Hardcovers at Amazon," *New York Times,* July 19, 2010.

32. Kathryn Zickuhr and Lee Rainie, "E-Reading Rises as Device Ownership Jumps," *PewInternet,* January 16, 2014.

33. Frank Catalano, "Traditional and Publishers' Ebook Sales Drop as Indie Authors and Amazon Take Off," *GeekWire,* May 19, 2018; Sian Cain, "Ebook Sales Continue to Fall as Younger Generations Drive Appetite for Print," *Guardian,* March 17, 2017.

34. Jim Milliot, "E-Book Sales Fell 10% in 2017," *Publishers Weekly,* April 25, 2018; Sian Cain, "Ebook Sales Continue to Fall."

35. Milliot, "E-Book Sales Fell 10%."

36. Amaranth Borsuk, *The Book* (Cambridge, MA: MIT Press, 2018), x.

37. Quoted in Tim Masters, "Man Booker Prize Won by Julian Barnes on Fourth Attempt," *BBC News,* October 9, 2011. Barnes won with *The Sense of an Ending.*

38. Borsuk, *The Book,* 237.

39. Bell and Gemmell, *Total Recall,* 182.

40. Jessica Pressman, "Bookwork and Bookishness: An Interview with Doug Beube and Brian Dettmer," in *Book Presence in a Digital Age,* ed. Kiene Brillenburg Wurth, Kári Driscoll, and Jessica Pressman (New York: Bloomsbury Academic, 2018), 60. See also Kiene Brillenburg Wurth, "Book Presence: An Introductory Exploration," in ibid., 1–23.

41. John T. Hamilton, "Pagina Abscondita: Reading in the Book's Wake," in ibid., 31.

42. Steven Hall, *The Raw Shark Texts* (Edinburgh: Canongate, 2007), 65. Further page references appear parenthetically in the text.

43. Laura U. Marks, *Touch: Sensuous Theory and Multisensory Media* (Minneapolis: University of Minnesota Press 2002), xiii.

44. Tegmark contributed to the "quantum suicide" thought experiment, a kind of quantum Russian roulette wherein an experimenter sits before a gun linked to a quantum particle and pulls the trigger. The gun fires only if the particle spins clockwise. In at least one of many possible parallel universes, the scientist will never die: "Every possible outcome happens, according to [the] Many Worlds [interpretation of quantum mechanics], but there'll be no chance for the scientist to know about the ones where he gets killed. . . . [A]fter 10 or so clicks he will be convinced that a reality exists where he can survive this process forever" (Michael Brooks, "Enlightenment in the Barrel of a Gun," *Guardian*, July 9, 2009). To read *The Raw Shark Texts* in light of this theory is to accept that while Eric Sanderson may be a man who dies during a psychological break in one universe, in another he is reunited with his lost love after defeating a conceptual fish.

45. See Jessica Pressman, "The Aesthetic of Bookishness in Twenty-First-Century Literature," *Michigan Quarterly Review* 48, no. 4 (2009): 475.

46. That books and paper incur such traces of human bodies via haptic storage may be one of the reasons why *The Raw Shark Texts* depicts many of the characters in the novel (including Eric) as being literally embodied by printed text; see Julia Panko, "'Memory Pressed Flat Into Text': The Importance of Print in Steven Hall's *The Raw Shark Texts*," *Contemporary Literature* 52, no. 2 (2011), 286–90.

47. N. Katherine Hayles, "Material Entanglements: Steven Hall's *The Raw Shark Texts* as Slipstream Novel," *Science Fiction Studies* 28, no. 1 (2011): 124.

48. Kiene Brillenburg Wurth, "Posthumanities and Post-Textualities: Reading *The Raw Shark Texts* and *Woman's World*," *Comparative Literature* 63, no. 2 (2011): 138.

49. Jacob Rabinow, "The Notched-Disk Memory," *Electrical Engineering* 71, no. 8 (1952): 746.

50. Jonathan Safran Foer, *Extremely Loud and Incredible Close* (London: Penguin, 2006), 1. Further page references appear parenthetically in the text.

51. Sarah Boxer, "One Camera, Then Thousands, Indelibly Etching a Day of Loss," *New York Times*, April 3, 2012.

52. Pasquale, *Black Box Society*, 3.

53. Dominic Head, *The State of the Novel: Britain and Beyond* (Chichester, UK: Wiley-Blackwell, 2008), 141.

54. Aaron Mauro, "The Languishing of the Falling Man: Don DeLillo and Jonathan Safran Foer's Photographic History of 9/11," *Modern Fiction Studies* 57, no. 3 (2011): 596.

55. Birgit Däwes, "On Contested Ground (Zero): Literature and the Transnational Challenge of Remembering 9/11," *American Studies* 52, vol. 4 (2007): 520.

56. Sonia Baelo-Allué, "The Depiction of 9/11 in Literature: The Role of Images and Intermedial References," *Radical History Review* 2011, no. 111 (2011): 188. Baelo-Allué argues that this is precisely the reason that "images play a very important role" in *Extremely Loud and Incredibly Close* (ibid.).

57. Bonnie Mak, *How the Page Matters* (Toronto: University of Toronto Press, 2011), 66.

58. Bell and Gemmell, *Total Recall,* 30.

59. Liedeke Plate, "How to Do Things with Literature in the Digital Age: Anne Carson's *Nox,* Multimodality, and the Ethics of Bookishness," *Contemporary Women's Writing* 9, no. 1 (2015): 106.

60. Ibid., 109; Kiene Brillenburg Wurth, "Re-vision as Remediation: Hypermediacy and Translation in Anne Carson's *Nox,*" *Image and Narrative* 14, no. 4 (2013): 26.

61. Anne Carson, *Nox* (New York: New Directions, 2010), 2.2, 8.1. Further location references appear parenthetically in the text. Although *Nox* is not paginated, Carson's commentary sections are numbered, so I use this system to cite quotations from those sections. When citing the location of other content, I refer to the commentary number that precedes it.

62. Plate, "How to Do Things," 99; Wurth, "Re-Vision as Remediation," 24.

63. Plate, "How to Do Things," 105.

64. *Nox* is also evidence that a book may provide consolation to the grieving by requiring its author to connect with other people to produce it. Sara Tanderup writes that "whereas [Carson's] original scrapbook was a personal project ('I made an epitaph'), the replica is part of a collective process ('as close as *we* could get')—involving others; that is, a publishing firm, an editor and not least Carson's partner, Robert Currie, who is credited in the work for assisting in 'the design and realization' of the book" ("Nostalgic Experiments: Memory in Anne Carson's *Nox* and Doug Dorst and J. J. Abrams' *S.*," *Image and Narrative* 17, no. 3 [2017]: 48).

65. Cathy Caruth, *Unclaimed Experience: Trauma, Narrative, and History* (Baltimore: Johns Hopkins University Press, 1996), 4, emphasis added.

66. Notable examples include Don DeLillo's *Falling Man* (2007), Frédéric Beigbeder's *Windows on the World* (2003), and Ian McEwan's *Saturday* (2005). For a critical treatment of the post-9/11 novel, see Richard Gray's *After the Fall: American Literature Since 9/11* (Edinburgh: Edinburgh University Press, 2001); Martin Randall's *9/11 and the Literature of Terror* (Chichester, UK: Wiley-Blackwell, 2011); and Kristiaan Versluys's *Out of the Blue: September 11 and the Novel* (New York: Columbia University Press, 2009).

67. N. Katherine Hayles, *Electronic Literature: New Horizons for the Literary* (Notre Dame, IN: University of Notre Dame Press, 2008), 169.

68. Mauro, "The Languishing of the Falling Man," 598.

69. Citing Foer as an example, Philippe Codde argues that creating fictionalized accounts of violent historical events that reverse or undo these events is a strategy common in fictional works by third-generation Holocaust survivors ("Keeping History at Bay: Absent Presences in Three Recent Jewish American Novels," *Modern Fiction Studies* 57, no. 4 [2011]: 673–93).

70. Ray Kurzweil, *The Singularity Is Near: When Humans Transcend Biology* (New York: Penguin, 2005), 13, 326.

71. Bell and Gemmell, *Total Recall,* 146.

72. Quoted in Adrienne Matei, "New Technology Is Forcing Us to Confront the Ethics of Bringing People Back from the Dead," *Quartz,* January 27, 2017.

73. Newton, "Speak, Memory."

74. David Parisi, *Archaeologies of Touch: Interfacing with Haptics from Electricity to Computing* (Minneapolis: University of Minnesota Press, 2018), 3.

75. Peter N. Miller, "How Objects Speak," *Chronicle of Higher Education,* August 11, 2014; Willard McCarty, "28.254 Digital and Material?," *Humanist Discussion List* 28, no. 254, August 11, 2014, https://dhhumanist.org.

76. John Durham Peters, *Speaking into the Air: A History of the Idea of Communication* (Chicago: University of Chicago Press, 2000), 269–70.

77. Ibid., 270. For grief experts' views on digital immortality, see Jillian D'Onfro, "Digital Immortality: How Technology Will Bring Loved Ones Back to Life," *NBC News,* December 16, 2016.

78. Peter James, *Host* (1993; reprint, New York: Villard, 1995), 454, emphasis added. Embodiment's impact on identity is also unfortunately evident in *Host*'s pervasive sexism: James's narrative consistently privileges the perspective of people in male bodies, reducing women to sexual objects and reproducing, through Spring's story, the misogynist cliché of the psychopathic and jealous seductress.

79. Jacques Derrida, *Archive Fever: A Freudian Impression,* trans. Eric Prenowitz (Chicago: University of Chicago Press, 1995), 99.

80. Ibid., 97.

81. James, *Host,* 37.

Chapter 5: Shelf Life

1. Robert Coover describes these platforms as "fragile and short-lived" and argues that the development of stable platforms is vital to the development of the field ("The End of Books," *New York Times,* June 21, 1992).

2. Stuart Moulthrop and Dene Grigar, *Traversals: The Use of Preservation for Early Electronic Writing* (Cambridge, MA: MIT Press, 2017), 3.

3. Christine L. Borgman, *Big Data, Little Data, No Data: Scholarship in the Networked World* (Cambridge, MA: MIT Press, 2015), 4.

4. On the persistence of digitally stored information, see Matthew G. Kirschenbaum, *Mechanisms: New Media and the Forensic Imagination* (Cambridge, MA: MIT Press, 2008), esp. chap. 1.

5. N. Katherine Hayles, quoted in Lisa Gitelman, "'Materiality Has Always Been in Play': An Interview with N. Katherine Hayles," *Iowa Journal of Cultural Studies* 2, no. 1 (2002): 10.

6. Planned obsolescence in design may include physical obsolescence (as in the cases of "limited functional life design," "design for limited repair," and "design aesthetics that lead to reduced satisfaction") or technological obsolescence (as a result of "design for fashion" or "design for functional enhancement through adding or upgrading product features") (Joseph Guiltinan, "Creative Destruction and Destructive Creations: Environmental Ethics and Planned Obsolescence," *Journal of Business Ethics* 89, supp. 1 [2009]: 20).

7. Jason Farman, "Repair and Software: Updates, Obsolescence, and Mobile Culture's Operating Systems," *Continent* 6, no. 1 (2017): 22.

8. See Joseph A. Schumpeter, *Capitalism, Socialism, and Democracy* (New York: Harper and Brothers, 1942).

9. Guiltinan, "Creative Destruction," 21.

10. Initially available only as an iOS app, *Arcadia* relates a number of interconnected stories that span three time periods. The app interface resembles the London Tube map, with each colored line representing one character's journey.

11. Allison Muri, "Twenty Years after the Death of the Book: Literature, the Humanities, and the Knowledge Economy," *English Studies in Canada* 38, no. 1 (2012): 117.

12. See Jerome Klinkowitz, "Fiction: The 1960s to the Present," *American Literary Scholarship* 2006, no. 1 (2006): 335–57; Jonathan Arac, "What Kind of History Does a Theory of the Novel Require?," *NOVEL* 42, no. 2 (2009): 190–95; and Georg Lukács, *The Theory of the Novel: A Historico-Philosophical Essay on the Forms of Great Epic Literature* (1920), trans. Anna Bostock (Cambridge, MA: MIT Press, 1971).

13. Klinkowitz cites "the pervasiveness of television" among the factors contributing "to the 'death of the novel' crisis" ("Fiction," 353). Arac addresses the issue directly, arguing that "the dual perspective of world history and media transformations makes clear the contours of mid-twentieth-century arguments about the 'death of the novel.' These were Western lamentations, and they had to do with the tremendous shift by which popular narrative passed from print to cinema and then to television" ("What Kind of History," 193).

14. See Jonathan Arac, "'This Will Kill That': A Provocation on the Novel in Media History," *NOVEL* 44, no. 1 (2011): 6–7.

15. Margaret-Anne Hutton, "Plato, New Media Technologies, and the Contemporary Novel," *Mosaic* 51, no. 1 (2018): 179; Jessica Pressman, "The Aesthetic of Bookishness

in Twenty-First-Century Literature," *Michigan Quarterly Review* 48, no. 4 (2009): 465–82.

16. Nicole Howard, *The Book: The Life Story of a Technology* (Baltimore: Johns Hopkins University Press, 2009), viii.

17. Ibid., 11, 5.

18. Amaranth Borsuk, *The Book* (Cambridge, MA: MIT Press, 2018), 38–39, 73.

19. Peter Stallybrass, "Books and Scrolls: Navigating the Bible," in *Books and Readers in Early Modern England: Material Studies*, ed. Jennifer Anderson and Elizabeth Sauer (Philadelphia: University of Pennsylvania Press, 2002), 44.

20. Borsuk, *The Book*, 17.

21. Robert Darnton, *The Case for Books: Past, Present, and Future* (New York: Public Affairs, 2009), xiv.

22. Naomi S. Baron, *Words Onscreen: The Fate of Reading in a Digital World* (New York: Oxford University Press, 2015), 12, 80–88.

23. Borsuk, *The Book*, 181. Borsuk also considers how books are vulnerable to loss and destruction, concluding that "ephemerality is . . . a condition shared by physical and digital books" (182).

24. Quoted in Nicholson Baker, *Double Fold: Libraries and the Assault on Paper* (New York: Random House, 2001), 28.

25. Ibid., 242, 243–44.

26. See Garrett Stewart, *Bookwork: Medium to Object to Concept to Art* (Chicago: University of Chicago Press, 2011).

27. Harriet Janis and Sidney Janis, "Marcel Duchamp, Anti-Artist" (1975), in *Marcel Duchamp in Perspective*, ed. Joseph Maschek (Cambridge, MA: Da Capo, 2002), 38.

28. See Claire Armitstead, "Can You Judge a Book by Its Odour?," *Guardian*, April 7, 2017.

29. William Blades, *The Enemies of Books* (1880; reprint, London: Elliot Stock, 1888).

30. Hans van der Hoeven and Joan van Albada, "Lost Memory: Libraries and Archives Destroyed in the Twentieth-Century" (Paris: UNESCO, 1996), 9.

31. Theodor H. Nelson, *Literary Machines: The Report on, and of, Project Xanadu Concerning Word Processing, Electronic Publishing, Hypertext, Thinkertoys, Tomorrow's Intellectual Revolution, and Certain Other Topics Including Knowledge, Education, and Freedom* (1980; reprint, Sausalito, CA: Mindful Press, 1987), version 87.1, 3/25. On the intentional destruction of books and libraries, see Rebecca Knuth, *Burning Books and Leveling Libraries: Extremist Violence and Cultural Destruction* (Westport, CT: Praeger, 2006); and Rebecca Knuth, *Libricide: The Regime-Sponsored Destruction of Books and Libraries in the Twentieth Century* (Westport, CT: Praeger, 2003).

32. Geoffrey Nunberg, "The Places of Books in the Age of Electronic Reproduction," *Representations* 42 (Spring 1993): 14–15.

33. The protagonist in Cormac McCarthy's *The Road* brings his son's book along on their grueling journey, but the child is "too tired for reading." In another scene, the protagonist recalls an earlier memory of standing "in the charred ruins of a library where blackened books lay in pools of water" ([New York: Vintage, 2006], 10, 187). In Jim Crace's *The Pesthouse,* "there are no books, no fragments of paper, [and] no literacy" (Caroline Edwards, "Microtopias: The Post-Apocalyptic Communities of Jim Crace's *The Pesthouse,*" *Textual Practice* 23, no. 5 [2009]: 777).

34. Kathleen Fitzpatrick, "The Exhaustion of Literature: Novels, Computers, and the Threat of Obsolescence," *Contemporary Literature* 43, no. 3 (2002): 521.

35. Ibid., 519.

36. Paul K. Saint-Amour, *Tense Future* (Oxford: Oxford University Press, 2015), 137–38, 139.

37. Ibid., 139–40.

38. Mary Lynn Ritzenthaler, *Preserving Archives and Manuscripts* (Chicago: Society of American Archivists, 2003), 25. According to Matija Strlič, the quality of paper has improved since the 1980s, when "the technology changed because of environmental concerns about the chlorinated chemicals emitted through the manufacture process. The happy consequence of that was that the paper became more stable again" (quoted in Armitstead, "Can You Judge a Book").

39. Virginia Woolf, *Orlando: A Biography* (1928), ed. Mark Hussey (New York: Harcourt, 2006), 208.

40. Ritzenthaler, *Preserving Archives and Manuscripts,* 25.

41. Robert Darnton, *The Case for Books,* 21–42, 37–38 (emphasis added), 38.

42. Bob Brown, *The Readies* (Bad Ems, Germany: Roving Eye, 1930), 31.

43. James Joyce, *Ulysses* (1922), ed. Hans Walter Gabler (New York: Vintage, 1986), 9.352–53, 13.1259, 3.141.

44. Filippo Marinetti, "The Founding and Manifesto of Futurism" (1909), in *Futurism: An Anthology,* ed. Lawrence Rainey, Christine Poggi, and Laura Wittman (New Haven, CT: Yale University Press, 2009), 52.

45. I have simplified the script that Melba Cuddy-Keane reproduced for publication, accepting insertions and removing deleted words. See Leonard Woolf and Virginia Woolf, "Are Too Many Books Written and Published?" (1927), *PMLA* 121, no. 1 (2006): 243.

46. Melba Cuddy-Keane, "Introduction," in Woolf and Woolf, "Are Too Many Books Written and Published?," 236–37.

47. Walter Benjamin, "The Storyteller" (1936), in *Illuminations: Essays and Reflections,* ed. Hannah Arendt, trans. Harry Zohn (New York: Schocken, 1968), 90.

48. Lisa Gitelman, *Always Already New: Media, History, and the Data of Culture* (Cambridge, MA: MIT Press, 2008), 18–19.

49. Shane Leslie, "*Ulysses*," *Quarterly Review* 238, no. 473 (1922): 234.

50. T. S. Eliot, "The Metaphysical Poets" (1921), in *The Waste Land and Other Writings* (New York: Modern Library, 2001), 232.

51. T. S. Eliot, "London Letter," *Dial* 72 (January–June 1922): 513.

52. Leonard Diepeveen, *The Difficulties of Modernism* (New York: Routledge, 2003), 55.

53. Ibid., 133.

54. Kenneth Goldsmith, *Uncreative Writing: Managing Language in the Digital Age* (New York: Columbia University Press, 2011), 12.

55. Michael Schaub, "*The Familiar* by Mark Z. Danielewski Review—What the Font Is Going On?," *Guardian,* May 12, 2015.

56. Quoted in Alexander Starre, *Metamedia: American Book Fictions and Literary Print Culture after Digitization* (Iowa City: University of Iowa Press, 2015), 197.

57. "Bells and whistles," *Oxford English Dictionary Online.*

58. Franco Moretti, *Graphs, Maps, Trees: Abstract Models for a Literary History* (London: Verso, 2005), 30. Ellipses in brackets are Moretti's.

59. Jonathan Culler, "Toward a Theory of Non-Genre Literature," in *Theory of the Novel: A Historical Approach,* ed. Michael McKeon (Baltimore: Johns Hopkins University Press, 2000), 52, 53, 54, 55.

60. See, for instance, Leah Price, *How to Do Things with Books in Victorian Britain* (Princeton, NJ: Princeton University Press, 2013); and Patrick Collier, *Modern Print Artefacts: Textual Materiality and Literary Value in British Print Culture, 1890–1930s* (Edinburgh: Edinburgh University Press, 2016).

61. A number of contemporary publishers have drawn attention to this relationship in order to playfully tease the book's obsolescence. The Argentinian publisher Eterna Cadencia released an anthology called *El Libro Que No Puede Esperar* (*The Book That Can't Wait*) (2012), printed with ink that disappears once it has been exposed to air and light for about two months (an experiment reminiscent of William Gibson's 1992 work *Agrippa: A Book of the Dead*). The rationale, writes Baron, borrowing a metaphor from Ray Bradbury, is "to light a fire under well-meaning readers who start reading a book but tend to lose track of the undertaking partway through" (*Words Onscreen,* 142).

62. See Jerome J. McGann and Lisa Samuels, "Deformance and Interpretation," *New Literary History* 30, vol. 1 (1999): 25–56.

63. Gretchen Henderson, *Galerie de Difformité* (Lake Forest, IL: Lake Forest College Press, 2011), 231.

64. John Lurz, *The Death of the Book: Modernist Novels and the Time of Reading* (New York: Fordham University Press, 2016), 55.

65. Joyce, *Ulysses,* 17.1361–1407.

66. Mark Z. Danielewski, personal communication.

67. I have verified that the Kim Forbes memorial webpage still exists, but I was not able to find the messages that McIntosh quotes either through a Google search or by searching or browsing the site's discussion forum.

68. Daniel Punday, *Writing at the Limit: The Novel in the New Media Ecology* (Lincoln: University of Nebraska Press, 2012), 151.

69. Baron found that readers in the United States, Germany, and Japan expressed a preference for print for pleasure reading, particularly in the case of longer texts (*Words Onscreen,* 80–92).

70. Reinhold Bauer, "Failed Innovations—Five Decades of Failure?," *Icon,* 20, no. 1 (2014): 37.

71. Erkki Huhtamo and Jussi Parikka, "Introduction: An Archaeology of Media Archaeology," in *Media Archaeology: Approaches, Applications, and Implications,* ed. Erkki Huhtamo and Jussi Parikka (Berkeley: University of California Press, 2011), 3.

72. See http://www.deadmedia.org.

73. Eric Kluitenberg, "On the Archaeology of Imaginary Media," in Huhtamo and Parikka, *Media Archaeology,* 51; Huhtamo and Parikka, "Introduction," 3.

74. Gitelman, *Always Already New,* 153.

75. Quoted in Ed Finn, "An Interview with Margaret Atwood," *Slate,* February 6, 2015.

76. Merve Emre, "This Library Has New Books by Major Authors, but They Can't Be Read Until 2014," *New York Times,* November 1, 2018.

77. Quoted in Mariann Enge, "In a Hundred Years All Will Be Revealed," *Kunstkritikk,* June 30, 2014.

78. Paul Benzon, "On Unpublishing: Fugitive Materiality and the Future of the Anthropocene Book," in *Publishing as Artistic Practice,* ed. Annette Gilbert (Berlin: Sternberg, 2016), 284. Although, as Benzon points out, the Future Library's emphasis on print and paper "elides the underlying digital backstory of the texts in question," the project's fetishizing of *print* textuality—including the forest ceremony where a printed manuscript is handed over to the library—not only imagines that the project will survive due to print media but also, by preserving the print manuscript, planting trees, and storing a printing press and printing materials, positions print as the more stable medium (285).

79. Jeffrey R. Di Leo, "Future Readers," *American Book Review* 36, no. 5 (2015): 2.

80. Benzon, "On Unpublishing," 285–286; Enge, "In a Hundred Years."

81. Bonnie Mak, *How the Page Matters* (Toronto: University of Toronto Press, 2011), 12.

82. Bob Schildgen, "Green Life: How Much Paper Does One Tree Produce?," *Sierra,* July 4, 2014.

83. "Codex," *Oxford English Dictionary Online.*

84. Daniel Goleman and Gregory Norris, "How Green Is My iPad," *New York Times,* April 4, 2010.

85. See, for example, Brett H. Robinson, "E-Waste: An Assessment of Global Production and Environmental Impacts," *Science of the Total Environment* 408, no. 2 (2009): 183–91.

86. Nicholson Baker, "A New Page: Can the Kindle Really Improve on the Book?," *New Yorker,* August 3, 2009.

87. "E-Books King: Stephen the First," *Wired,* March 14, 2000.

88. Stephen King, *UR* (Amazon Digital Services, 2009), loc. 168. Further location references will appear parenthetically in the text.

89. Stephen King, "Stephen King on the Kindle and the iPad," *Entertainment Weekly,* March 26, 2010.

90. As Starre defines this phrase, "a medial mise en abyme . . . occurs whenever the artwork activates its own storage and reproduction medium" (*Metamedia,* 155).

91. Sebastian Schmieg, with Silvio Lorusso, *56 Broken Kindle Screens,* http://www.sebastianschmieg.com.

92. Roger Chartier, "Languages, Books, and Reading from the Printed Word to the Digital Text," trans. Teresa Lavender Fagan, *Critical Inquiry* 31, no. 1 (2004): 142.

93. D. H. Lawrence, "Review [of *Manhattan Transfer,* by John Dos Passos]" (1927), in *John Dos Passos,* ed. Barry Maine (New York: Routledge, 1988), 71.

94. See Jessica Pressman, *Digital Modernism: Making It New in New Media* (Oxford: Oxford University Press, 2014); and Goldsmith, *Uncreative Writing.*

95. See, for example, Douglas Mao and Rebecca L. Walkowitz, "The New Modernist Studies," *PMLA* 123, no. 3 (2008): 737–48; and Susan Stanford Friedman, *Planetary Modernisms: Provocations on Modernity across Time* (New York: Columbia University Press, 2015).

96. See Douglas Mao and Rebecca L. Walkowitz, *Bad Modernisms* (Durham, NC: Duke University Press, 2006); and Paul K. Saint-Amour, "Weak Theory, Weak Modernism," in *Modernism/modernity* 25, no. 3 (2018): 437–59.

97. See Collier, *Modern Print Artefacts,* intro.

98. Douglas Mao, "Our Last September: Climate Change in Modernist Time," in *The Contemporaneity of Modernism: Literature, Media, Culture,* ed. Michael D'Arcy and Mathias Nilges (New York: Routledge, 2016), 33–34.

99. Virginia Woolf, "Character in Fiction" (1924), in *Virginia Woolf: Selected Essays,* ed. David Bradshaw (Oxford: Oxford University Press, 2008), 51.

100. See Saint-Amour, *Tense Future;* and Mao, "Our Last September."

101. Mark Goble, *Beautiful Circuits: Modernism and the Mediated Life* (New York: Columbia University Press, 2010), 318.

102. Jacques Derrida, "The Book to Come," in *Paper Machine,* trans. Rachel Bowlby (Stanford: Stanford University Press, 2005), 16–17.

INDEX

hypermediacy; transparent
mediation

Guillory, John, 52, 254n89

Hall, Steven, and *The Raw Shark Texts*:
books and paper in, 178, 180–81,
183–84; digital media in, 179–80;
haptic storage in, 166, 178, 183–86;
metamedial aspects of, 18, 177–78;
representation in, 180–83; typography in, 179

handwriting: aura of, 156–57; in
novels: 184, 191, 195–96; Virginia
Woolf's, 151, 271n85

haptic storage, 30–31, 150–62, 183–86,
272n91; in Hall's *The Raw Shark
Texts*, 178, 183–86; in Virginia
Woolf's *Orlando*, 143, 150–59. *See
also* book: metaphysics of presence; book: tactility of; interface

Hayles, N. Katherine, 5, 27, 98, 242n12,
245n45, 246n61; on Danielewski's
Only Revolutions, 107, 110; on flickering signifiers, 95, 203, 261n61;
on Foer's *Extremely Loud and
Incredibly Close*, 195–96; on Hall's
The Raw Shark Texts, 184; on hyper
attention, 16, 244n40; on information's embodiment, 171, 174

Henderson, Gretchen, and *Galerie de
Difformité*, 222

Henry, Edward A., 129

Hong, Cathy Park, 122

Howard, Nicole, 207

Huhtamo, Erkki, 225

Hunter, J. Paul, 9, 74–75, 115, 242n14

Huxley, Aldous, 39

hyper attention, 16, 45, 80–81, 244n40

hypermediacy: of books, 18–20, 32,
73–77, 82–83, 95–96, 104, 124–26,
197–98; defined, 5, 241–42n8; in

Internet search, 121; of modernist typography, 140–41. *See
also* Bolter, Jay David; book:
metaphysics of presence; Grusin,
Richard; novel, metamedial:
hypermediacy of; transparent
mediation

index cards, 52–55, 112, 161. *See also*
card catalogues; Mundaneum;
Otlet, Paul

information: aesthetics of, 4–6, 46,
62, 85–86; defined, 11–12, 29; as
immaterial, x–xi, 3–4, 20, 124–
26, 138–40, 159–61, 168, 170–74,
176–77; made meaningful by
form (*see* formation of data); as
neutral, 7, 68–70, 86–87, 91, 121.
See also data; novel: informational
aspects of; modern information
management

information, metaphors of: as black
box, 4, 82, 91, 95–96; as circle and
sphere, 85, 110; as cloud, 4, 82, 85–
86, 91; as flood, 84; human mind
as, 164–65, 173–74. *See also* book:
and bodies

information management. *See* modern information management

information scale, 4, 14–18, 25–26, 38;
in book production, 20, 26, 37–39;
in digital data, 14–16, 79–83, 87,
169–70; as ineffable, 20, 84–86, 88,
91–92, 113–14; in online information, 30, 115, 120–21; in periodical
and newspaper production, 26,
41–42. *See also* Big Data; information shock; Internet: scale of;
modern information management; novel: information scale
represented in

McGann, Jerome, 141, 222, 246n53, 264n104, 268n41

McIntosh, Matthew, and the*Mystery .doc*, 30, 82–83, 120–22, 265n121; as book-object, 96–97, 99, 101–3; fiction vs. autobiography in: 97, 99, 103; genre of, 114, 116–18, 119–20; Internet search in, 99–104, 223; as maximalist, 96–99, 102; metamedial aspects of, 102–3; typography in, 100, 102; as unreadable, 102. *See also* maximalist novel

McKeon, Michael, 12, 72

media archaeology, 224–25

media ecology, xi, 239–40; print book as, 22

media history, 24, 28–29, 121, 224–26, 237–40

medial ideology: defined, 139, 172; of microform, 138–40, 159–61. *See also* information: as immaterial

mediation, ix–xi, 4–8, 14–15, 18–21. *See also* algorithmic mediation; computational mediation; hypermediacy; interface; modern information management; novel, metamedial; transparent mediation

metafiction, 19, 98–99. *See also* novel, metamedial

metamedial novel. *See* novel, metamedial

microfiche, 127, 128, 130. *See also* microform

microfilm, 127, 134, 140, 209–10. *See also* Brown, Bob: Readies project; microform

microform: vs. books, 128–30, 133, 138–40, 161–62, 213–14; history of, 127–29, 265–66n8; medial ideology of, 138–40, 159–61; reading machines, 132–33, 138–39, 268n47;

revolutionary potential of, 129–30, 137; use of screens, 20, 26, 31, 124–25, 130. *See also* Brown, Bob: Readies project; Fiske, Bradley A.; medial ideology

mimesis, 1–2, 150–51, 174–75, 177, 180–84, 218. *See also* representation

Mirrlees, Hope, and *Paris,* 142, 271n87

modern epic. *See under* Moretti, Franco

modern information management, x–xi, 3–4, 7, 20–21, 26, 55–57. *See also* hypermediacy; information scale; interface; systematic management; transparent mediation

modern information mediation. *See* modern information management

modernism: difficulty of, 64, 217–18 (*see also* unreadability); experiments with media form, x, 141–43; fragmentation in, 49, 62–63, 76; media culture of, 7, 24–28, 35, 37–39, 41–44, 75–77, 159–62; parallels with contemporary media culture, 4, 7, 24–26, 28–29, 120, 132, 212, 225, 236–39; parallels with contemporary metamedial novels, 5–7, 10–11, 24–28, 102, 120, 236–39. *See also* book: death and future of; Brown, Bob: Readies project; encyclopedic narrative: in modernism; information shock

modernist studies, 25, 27, 32, 236–37, 247n65

monographic principle, 51–54, 62, 76

Montfort, Nick, 202, 275n25

Moretti, Franco: on distant reading, 80; on modern epic, 58, 253–54n83; on the novel, 221

Moulthrop, Stuart, 202–3